NUTRITION
At A Glance

THE AUTHOR

Dr. Madhvi Awasthi is a 'Nutrition and Dietetics' expert and currently serving as a 'Clinical Nutritionist'. Dr. Awasthi has worked as an Assistant Professor at Lovely Professional University, Punjab. She has presented various papers in National and International conferences.

NUTRITION
At A Glance

Dr. Madhvi Awasthi

2018

Daya Publishing House®

A Division of

Astral International Pvt. Ltd.

New Delhi – 110 002

Cataloging in Publication Data--DK
Courtesy: D.K. Agencies (P) Ltd. <docinfo@dkagencies.com>

Awasthi, Madhvi, author.
Nutrition : at a glance / Dr. Madhvi Awasthi.
pages cm
Includes bibliographical references.

ISBN 9789388173063 (Int. Edition)

1. Nutrition. 2. Food. I. Title.
LCC RA784.A93 2018 | DDC 613.2 23

Published by : **Daya Publishing House®**
A Division of
Astral International Pvt. Ltd.
– ISO 9001:2015 Certified Company –
4736/23, Ansari Road, Darya Ganj
New Delhi-110 002
Ph. 011-43549197, 23278134
E-mail: info@astralint.com
Website: www.astralint.com

Digitally Printed at : **Replika Press Pvt. Ltd.**

Preface

'Nutrition at A Glance' is acknowledged as an emerging field in the bio-medical sciences exploring the role of food in treating diseases and to provide balanced mental and physical health.

The quick review provided in this book enables the reader and specifically, the students of 'Nutrition and Dietetics' and allied sciences about the quick facts of this emerging field. The information given in this book has been extracted from valuable and authenticated sources and a sincere effort has been made to present the facts in brief for easy grasping and understanding by the reader. The book also contains advanced facts and detailed explanation of important terms which will be of immense help for UG & PG students during their comprehensive exams, interviews and practical viva.

The book also includes important facts about the field of Food Science and Technology because knowledge of technical advances in food industry and development of newer food products is mandatory for the students of Nutrition and Dietetics to be in sync with the latest happenings and for excelling in their respective fields.

Dr. Madhvi Awasthi

Contents

1

Nutrition: An Introduction

The basic requirement of human life is oxygen, then water and after this food. Food is anything which we eat and drink and which gives us nutrients. The study of these nutrients is done under a field known as Nutrition. 'Nutrition' technically can be defined as 'A branch of science that deals with ingestion, digestion, transportation and absorption of nutrients and finally excretion of waste in form of feces and urine'.

Functions of food

- **Physiological functions of food:** The basic and foremost aim of food is to cater for nutritional requirements of the body and to maintain requisite energy levels, so as to sustain the body in adversity. In another sense, physiological function of food is nothing but the bodily requirement of growing and sustaining.
- **Social functions of food:** The basic physiological need of food developed the sense of togetherness among the pre-historic humans to gather and share the food. Thus, the food itself has acted as a stepping stone for formation of society. Food has always been the central part of any community and evokes strong regional and emotional feelings. In addition, food has been integral part of our social, cultural and religious life.

- **Psychological functions of food:** Food plays a vital aspect in life of human beings. In addition of satisfying physical and social needs, it evokes strong emotional feelings and provides a sense of security and emotional bondage. For example, a student living far from home in alien atmosphere longs mostly for home cooked food.

Classification of Food

1. As per nutrient density and functions

i. **Energy yielding foods:** Foods rich in carbohydrates and fats are called energy yielding foods. They provide energy to sustain the involuntary processes essential for continuance of life, to carry out various professional, household and recreational activities and to convert food ingested into usable nutrients in the body. The energy needed is supplied by the oxidation of foods consumed. *e.g.* Cereals, roots and tubers, dried fruits, oils, butter and ghee.

ii. **Body building foods:** Foods rich in protein are called body building foods. Milk, meat, eggs and fish are rich in proteins of high quality. Pulses and nuts are good sources of protein but the protein is not of high quality. These foods help to maintain life and promote growth. They also supply energy.

iii. **Protective and regulatory foods:** Foods rich in protein, minerals and vitamins are known as protective and regulatory foods. They are essential for health and regulate activities such as maintenance of body temperature, muscle contraction, maintaining water balance, clotting of blood, removal of waste products from the body and maintaining heartbeat. Milk, egg, liver, fruits and vegetables are protective foods.

2. ICMR Five Food Groups

Table 1.1: Five Food Group System

Food Group	*Main Nutrient*
I. Cereals, Grains and Products :	
Rice, Wheat, Ragi, Bajra, Maize, Jowar, Barley, Rice flakes, Wheat flour	Energy, Protein, Invisible fat Vitamin B_1, Vitamin B_2, Folic Acid, Iron, Fibre.

Food Group	*Main Nutrient*
II. Pulses and Legumes :	
Bengal gram, Black gram, Green gram, Red gram, Lentil (whole as well as dhals) Cowpea, Peas, Rajmah, Soyabeans, Beans.	Energy, Protein, Invisible fat, Vitamin B_1, Vitamin B_2, Folic Acid, Calcium, Iron, Fibre.
III. Milk and Meat Products :	
Milk : Curd, Skimmed milk, Milk, Cheese	Protein, Fat, Vitamin B_{12}, Calcium.
Meat : Chicken, Liver, Fish, Egg, Meat.	Protein, Fat, Vitamin B_2
IV. Fruits and Vegetables :	
Fruits :	Carotenoids, Vitamin C, Fibre.
Mango, Guava, Tomato Ripe, Papaya, Orange. Sweet Lime, Watermelon.	Invisible Fats, Carotenoids, Vitamin B_2.
Vegetables (Green Leafy) :	Folic Acid, Calcium,
Amaranth, Spinach, Drumstick leaves, Coriander leaves, Mustard leaves, fenugreek leaves .	Iron, Fibre.
Other Vegetables :	
Carrots, Brinjal, Ladies fingers, Capsicum, Beans, Onion, Drumstick, Cauliflower.	Carotenoids, Folic Acid, Calcium, Fibre
V. Fats and Sugars :	
Fats :	Energy, Fat, Essential Fatty Acids
Butter, Ghee, Hydrogenated oils, Cooking oils like Groundnut, Mustard, Coconut.	
Sugars :	
Sugar, Jaggery	Energy

Source : Gopalan C, Rama Sastri BV and Balasubramaniam SC. 1989. Nutritive Value of Indian Foods. National Institute of Nutrition, ICMR, Hyderabad.

Significance of the five-food group system

The five food group system can be used for the following purposes :

1. Planning wholesome balanced menus to achieve nutritional adequacy.
2. Assessing nutritional status: A brief diet history of an individual can disclose inadequacies of food and nutrients from any of the five groups.

Food Pyramid

The food guide pyramid was introduced in 1992 by USDA (United States Department of Agriculture) as a general plan of what to eat each day. The food guide pyramid is a valuable tool for planning a health promoting diet. By incorporating the principle of balance, variety and moderation, an individual can still eat their favorite foods while following the food guide pyramid.

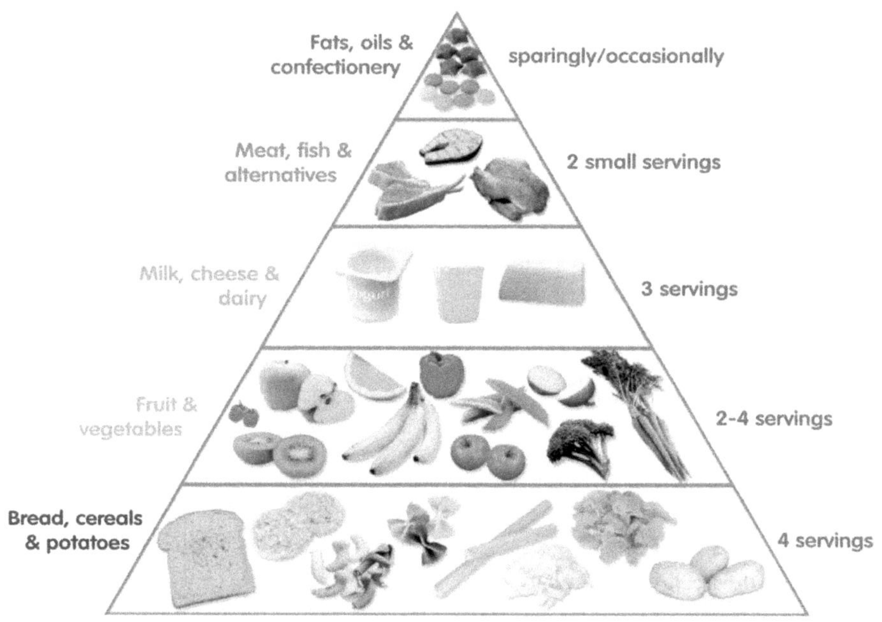

Fig 1.1: Food guide pyramid

Balanced Diet

The diet which provides all the nutrients to the body in required amount. A good or adequate diet is known as 'balanced diet'. A balanced diet yields daily nutrients in the proper amounts and proportion required by the body. A balanced diet must supply enough food to the body for deriving energy and enough protein for building up our tissue in various parts of our body. It must supply enough protein for building up our tissue in various parts of our body. Fats and oils are other major nutrients in a balanced diet. A balanced diet provides enough vitamins and minerals from natural sources.

Table 1.2: Balanced diet for adults-sedentary/moderate/heavy activity (number of portions)

Food Groups	Portion	Type of work					
		Sedentary		Moderate		Heavy	
		Male	Female	Male	Female	Male	Female
Cereals and millets	30	14	10	16	12	23	16
Pulse	30	2	2	3	2.5	3	3
Milk ml	100	3	3	3	3	3	3
Roots & tubers	100	2	1	2	1	2	2
Green leafy vegetables	100	1	1	1	1	1	1
Other vegetables	100	1	1	1	1	1	1
Fruits	100	1	1	1	1	1	1
Sugar	5	5	4	8	5	11	9
Fats and Oils (visible)	5	4	4	6	7	11	8

For non-vegetarians substitute one pulse portion with one portion of egg/meat/chicken/fish.
For infants introduce egg/meat/chicken/fish around 9 months
Specific recommendations as compared to a sedentary woman:
Children:
1-6 years: ½ to ¾ the amount of cereals, pulses and vegetables and extra cup of milk.
7-12 years: extra cup of milk
Adolescent girl: extra cup of milk
Adolescent boy: Diet of sedentary man with extra cup of milk

Source: Dietary gudielines for Indians—A manual 1999 National Institute of Nutrition. ICMR, Hyderabad 500 007.

Classification of nutrients

Nutrients can be classified on the basis of requirement:

Macronutrients: These are those nutrients which are required in grams. *e.g.* Carbohydrates, fat and proteins.

Micronutrients: These are those nutrients which are required in small quantities in milligrams or micro grams. Eg. Minerals and vitamins.

If human body lacks any of these nutrients or take them in excess we suffer either from deficiency or excess. This state is known as malnutrition. If one is taking less amount of nutrients than prescribed by or other health governing bodies then he/she will suffer from under nutrition (Table 1.3). And if one is taking excess amount of nutrients than prescribed then he/she will suffer from over nutrition.

Table 1.3: Indicators of Under nutrition

Indicator	*Interpretation*	*Comments*
Stunting	Low height for age	Chronic malnutrition
Wasting	Low weight for height	Acute malnutrition
Underweight	Low weight for age	Both acute & chronic malnutrition

Antioxidants

Antioxidants are compounds that protect biological systems against the potentially harmful effects of processes or reactions that can cause extensive oxidations. The ultimate purpose of food antioxidants is to inhibit oxidative reactions that cause deterioration of quality (e.g., of flavor, colour, nutrient composition, texture). They prevent the deterioration of food lipids by free radical scavenging, inactivation of peroxides and other reactive oxygen species, chelation of metals and quenching of secondary lipid oxidation products that produce rancid odours.

Phytochemicals

They are non-nutrient compounds found in plant derived foods and have biological activity in the body. These chemicals of plant origin include terpenes, phytosterols, flavanoids, theols and allylic sulphides which are antimutagenic and anticarcinogenic agents and thus render nutraceutical properties. More than 5000 phytochemicals have been identified in plant foods and many more remain to be discovered. Carotenoids are phytochemicals that are abundantly present in yellow-orange coloured fruits and vegetables and their increased consumption is associated with several health benefits. Nuts contain many phytochemicals (flavonoids, phenolic compounds, isoflavones, sterols *etc.*), which have been shown to be inversely related to CHD.

Neutraceutical

In the year 1989, the term "Nutraceutical," was coined by Dr. Stephen De Felice, a physician who founded the Foundation for Innovation in Medicine, USA. At that time, Dr. De Felice defined "Nutraceutical" as "Any food or parts of a food that provides medical or health benefits, including the prevention and treatment of diseases". Nutraceuticals are the health promoting compounds or products that have been isolated or purified from food sources and they are generally sold in a medicinal (usually pill) form. A good example is a group of compounds called isoflavones that are isolated from soybean seeds and packaged into pills that women can use instead of synthetic compounds during hormone replacement therapy. Other examples of nutraceutical products include fish oil like cod liver oil capsules, herb extracts, glucosamine, lutein-containing multivitamin tablets and antihypertensive pills that contain fish protein-derived peptides.

2

Energy in Human Nutrition

Introduction

We know, energy is the ability, or power to do work. Being a student of nutrition, you may know that we derive energy from carbohydrates, protein and fats which are called as macronutrients. We require energy for many functions like walking, running breathing, chewing and even for digesting food. We need energy to maintain our body temperature, keep our heart beating and it enables us to understand what we read.

Key notes on Energy

- ★ One Kcal is equal to 4.186 KJ.
- ★ One gram of carbohydrates give 4.1 Kcal of energy in bomb calorimeter and 4 Kcal in human body on metabolism.
- ★ One gram of proteins provide 5.65 Kcal of energy in bomb calorimeter and 4 Kcal in human body on metabolism.
- ★ One gram of fats give 9.45 Kcal of energy in bomb calorimeter and 9 Kcal in human body on metabolism.
- ★ One gram of alcohol provides 7 Kcal.
- ★ One gram of short chain fatty acid give 6-7 Kcal.
- ★ Loss of energy given by carbohydrates, proteins and fats during metabolism in human body is 2, 5 and 8 per cent, respectively.

* One grams of protein can hold up to 4 g of water.
* One gram of fat can only hold 0-2 g of water.
* Mitochondria is known as power house of living cells.
* 1 ATP gives 7Kcal.
* The various components of energy expenditure which are taken into account to determine the Energy Requirements (ER) includes BMR, Total Energy Expenditure (TEE), energy for physical activity and energy for deposition in growing tissues during infancy, childhood and adolescence.
* The factors that affect BMR are: Body composition, gender, age, body size and surface area, sleep, body temperature, pregnancy, environmental temperature and smoking.
* During sleep BMR is 10 per cent low.
* With every 1°C rise in temperature, BMR is increased by 13 percent.
* Green revolution came in to existence in 1960.
* The 7th April is celebrated as World Health Day.
* Green revolution stands for crop improvement in India, blue for fish production, white for milk, yellow for oilseeds and brown for non conventional energy sources.
* GOBI is Growth Monitoring, Oral Rehydration, Breast Feeding and Immunization.

Methods of Energy Measurement

1. Indirect methods:

* Bomb calorimeter
* Open circuit indirect calorimeter
* Automatic respirometer
* Physical activity diary method
* Caltract activity computer
* Wireless heated machine
* Nutriguide computer programme
* Hallow scale respirometer

- ★ Infra red carbon dioxide indicator
- ★ Oxygen monitor

2. Direct method:

- ★ By measuring energy released by human body in a closed controlled chamber.

Units of Energy – Calorie and Joule

- ★ The unit of energy, kilocalorie (kcal) was used for a long time. Recently the International Union of Sciences and International Union of Nutritional Science (IUNS) have adopted 'Joule' as the unit of energy in place of kcal. These units are defined as follows.
- ★ A joule is defined as the energy required to move 1 kg mass by 1 meter by a force of 1 Newton acting on it.
- ★ One Newton is the force needed to accelerate 1 kg mass by less than a second.
- ★ Kcal is defined as the heat required to raise the temperature of 1kg of water by 1°C. (From 16.5°C to 17.5°C).

1 Kcal = 4.184 Kj (Kilo Joules)

1000 Kcal = 4184 = 4.18 Mj (mega joules)

1 Kj = 0.239 Kcal.

Energy Balance

- ★ Energy balance is the relationship between energy intake and energy expenditure.
- ★ To maintain daily energy balance the total energy requirement of a human being is the number of kcal required to restore daily basal metabolic loss and loss from exercise and other physical activities.
- ★ Energy equilibrium occurs when the calories consumed from food source (energy intake) is equal to amount of energy expended.
- ★ Positive energy balance results, if energy intake exceeds energy expended. The excess energy consumed is stored, resulting in weight gain.
- ★ Positive energy balance is desired during the growth stages of the life cycle (pregnancy, infancy, childhood & adolescence) and it

instead restore body weight to healthy level after losses caused by starvation, disease or injury.

- ★ During other times such as adulthood, positive energy balance over time can cause body weight to climb to unhealthy level.
- ★ The process of aging itself does not cause weight gain rather weight gain stems from a pattern of excess food intake coupled with limited physical activity and slower metabolism.
- ★ Negative energy balance results. when energy intake is less than energy expenditure.
- ★ Weight loss occurs because energy stored in the body in the form of fat or glycogen in liver and muscle is used to make up for the shortfall in energy intake.
- ★ Negative energy balance is desired in adults when body fatness exceeds healthy level.
- ★ Negative energy balance during growth stages of the life cycle generally is not recommended because it can impair normal growth.

Table 2.1: Energy Cost of some Common Activities in Terms of BMR Units

Activity	*Energy Cost of Activities in BMR Units*
1. Sitting quietly	1.2
2. Standing quietly	1.4
3. Sitting at desk	1.3
4. Walking (3MPH)	3.7

Source: ICMR. 2002. Nutrient requirements and recommended dietary allowances for Indians, NIN.

3

Water and its Functions

Water: An Introduction

We cannot survive for long without water although we can survive without food for a quite few days. Water gives form and structure to cells and the body. It is also necessary for flushing out toxic substances. Approximately, 55 to 70 per cent of total body weight is made up of water. The percentage of water tends to decrease as a person gets older. Thus, infants and children have much higher content of water than adults. A fatty person has less water than a lean one. Water is an essential nutrient and is as important as oxygen to sustain life. During growth, water is also required for the formation of new tissues.

Functions

Water acts as a dissolving medium, building material, lubricant and temperature regulator as described below:

- Water is a universal solvent and is able to dissolve all the products of digestion.
- Water is a constituent of all body fluids, so it helps in the transportation of digested products to the appropriate organs *e.g.* blood which contain 90 per cent of water, carries carbon dioxide to the lungs, transport nutrients to the cells and waste nitrogenous material and salt to the kidneys.

- ★ Urine, which contains 97 per cent water has all the waste material dissolved in it and the body thus, is able to excrete soluble waste products of metabolism.
- ★ Water acts as a building material. It is used in construction of every cell. The cells of tissues of the body differ in their water content *e.g.* fatty tissues contain 20 per cent, bones 25 per cent and muscles 60 to 80 per cent.
- ★ Water serves as a lubricant in the joints and between internal organs.
- ★ It keeps the cells of the body moist and permits the passage of substances between the cells of blood vessels.
- ★ Water helps in removal of heat from the body by the evaporation of water from the skin surface. It is an effective way to eliminate the body heat.
- ★ The body is composed of 50-75 per cent of water, depending on age and body fatness. Water is essential for living, significantly more so than food.
- ★ Death will usually occur after one week without any liquids (2 to 3 days in the heat) but humans have been known to survive some months without food.
- ★ The common rate of sweat loss is about 275 ml in the heat. Prolonged exercise/performance can be impaired by depletion of the body's energy stores and by disturbances of water and electrolyte balance.
- ★ Exercise can exceed water loss up to two litres. Even more modest and common rates of sweat loss (*e.g.*, 1 - 1.5 l/h) sustained for 30 to 60 minutes can result in dehydration that is of sufficient magnitude to negatively affect cardiovascular, thermoregulatory and performance responses.
- ★ There is, however, a limit to the body's ability to maintain physiological homeostasis during exercise in a warm environment when the body is no longer capable of coping with the demands placed upon it.

The Relationship of Water and Exercise

In Exercise physiology, water plays an important role in two aspects: "Heat Dissipation" and "Physiological Homeostasis".

- ★ Heat Dissipation: Regulates body temperature through perspiration.
- ★ Physiological Homeostasis: Water helps to maintain the physical equilibrium to regulate its internal environment to maintain a stable constant condition for adapting the physical activity.
- ★ Normal body water turnover in an adult is from 1.7 to 2.3 L/day
    - Respiration: vigorous exercise can dissipate 1-5ml water per minute
    - Sweat: about 500ml per day (depending upon temperature and humidity)
    - Urine excretion: about 1000-1500ml per day
    - Stool excretion: about 100-200 ml per day

Effects of Dehydration on Physiology and Performance

Inadequate drinking during strenuous activity can result in dehydration and impaired athletic performance. During intense physical activity, athletes often do not drink enough fluids to replace what they lose in exercise, resulting in what's called "voluntary dehydration". The followings are the physiological consequences between difference dehydration levels:

- Fluid loss of as little as 1 per cent of body weight (about 600ml) cause heart rate to be elevated, cardiac output to be declined and core temperature to be raised.
- Fluid loss of 2 per cent of body weight (less than 4 pounds in a 200-pound athlete), due to the dysfunction of homeostasis can impair performance by raising the body temperature and increasing fatigue which fails to replace body fluids.
- Fluid loss of 4-6 per cent of body weight (about 2400-3600 ml) will decrease muscle endurance and heat spasm often strength and muscle endurance. In such cases, heat spasm often develops. The performance ability may lower by 20-30 per cent and also affects the anaerobic capacity. Besides, there are progressive increase in the concentrations of sodium and other dissolved substances in the blood plasma, a lower blood flow to the skin and a higher core temperature which will increase the chance of heat stroke.
- Fluid loss of over 6 per cent of body weight (> 3600ml) leads to symtoms of impaired cardiovascular function (*e.g.*, syncope,

heat exhaustion, heat spasm) and in severe cases, neurological failure and severe hyperthermia ensue (*e.g.* heat stroke).

Suggestions for Fluid Intake

- ★ Voluntary drinking before the thirst signals. Initially, one experiences thirst and discomfort when they lose 1 per cent of body fluid. If you feel thirsty, you are already dehydrated.
- ★ Try to avoid caffeinated and alcoholic beverages, while they do supply water to the body initially, contain diuretics that cause the body to lose water.
- ★ Cold beverages are more palatable during and after exercise and this greater palatability will increase fluid consumption by athletes.
- ★ Drinking cold beverages (8-12°C) causes a slight transient cooling of the upper digestive tract.
- ★ Sports drinks are intended to replenish electrolytes, sugar, water and other nutrients and are usually isotonic (containing the same proportions as found in the human body).
- ★ Non-athletes who use sports drinks should also be aware that sports drinks for athletes typically contain high levels of carbohydrates which will result in weight gain if consumed without a corresponding increase in exercise activity.
- ★ Try to avoid concentrated juice, sweetened drinks, chocolate milk and soda. These sugared beverages will lower the rehydration rate.
- ★ The choice of drink will depend on whether you need a drink to replace fluid losses or to provide more energy/carbohydrate or both. Either plain water or sports drink which contain 4 to 8 per cent carbohydrate are suitable.

Fluid Intake before Exercise

- ★ Prehydrating
  - Drink approximately 600ml of fluids up to two hours before an endurance exercise session.
  - Drink 200 ml of fluids 15 minutes before exercise.

Fluid Intake during Exercise

- ✯ During intense and prolonged exercise sessions, or exercising in a hot/humid environment, drink 150-350 ml of fluids after every 15-20 minutes (depending on individual sweating rate & exercise duration).
- ✯ Splashing some water onto the skin surface can help to reduce the sweat evaporation during exercise.
- ✯ Plain water is useful in non-endurance events of low intensity, where carbohydrate replacement is not the priority. Since it is much faster to be absorbed into the body.
- ✯ If exercise duration is longer than one hour, consuming sports drinks can sustain fluid-electrolyte balance and exercise performance.
- ✯ Ingestion of approximately 30-60 g of carbohydrate during each hour of exercise will generally be sufficient to maintain high rates of oxidation of blood glucose late in exercise and to delay fatigue.

Fluid Intake after Exercise

- ✯ Exercise, drink enough fluids to quench your thirst plus extra.
- ✯ Use weighing machine to check body weight after exercise
- ✯ Use body weight after exercise as a guideline. Drink 480ml of fluids for every pounds lose.

Excessive Fluid Consumption after Exercise

- ✯ Over-drinking 1-2L will increase the blood circulation, so as to increase the workload for the cardiovascular system.
- ✯ The potential danger of excessive fluid consumption which may, in the extreme, result in a low blood sodium concentration or hyponatremia, while the sodium concentration in the plasma falls below 135 mmol/L.
- ✯ Hyponatremia is a rare occurrence, it is a dangerous condition that may arise when athletes drink too much water, diluting the body's sodium.

Recommended Dietary Allowance (RDA)

- ★ The amount of water needed by an individual depends on many factors such as environmental temperature, humidity, occupation and the diet.
- ★ A person may need to drink about 1.5 to 2 liters of water per day.

Sources

- ★ Almost all the fruits and vegetables contain 70 to 95% water.
- ★ All food grains contain some amount of water. Cooked food also contains water.
- ★ Beverages and drinking water are good source of water for the body.

4

Protein in Human Nutrition

Proteins are the body building material chiefly composed of carbon, hydrogen, oxygen and nitrogen and some times sulphur. The major elements in proteins is nitrogen which may be present up to 14 to 20 per cent. Nutritional requirements of proteins and amino acids vary with age. Deficiency of proteins lead to protein energy malnutrition. Proteins are vital for human body growth and their important functions are as elaborated hereunder.

Functions of proteins

Proteins play their role in several crucial ways in body metabolism and in the formation of body structure.

Producing vital body structures

The primary function of protein is to give structural support to body cells and tissues. Each cell in the body contains proteins in the cell membrane that surrounds the cell and holds it together, in the organelles in the cytoplasm and in the nucleus itself where the DNA is housed.

Transporting nutrients

- ★ Many proteins function as transporters for other nutrients, carrying them through the bloodstream to cells and across cell membranes to sites of action. For example, lipoproteins transport large lipid molecules from small intestine, through the lymph and blood to body cells, hemoglobin carries oxygen from the lungs to cells.

* Few minerals and vitamins also have protein carriers that aid in their transport into and out of tissues and storage proteins. This includes transferrin and ferritin (carrier and storage proteins for iron), ceruloplasmin (a carrier protein for copper) and retinol- binding protein (a carrier protein for vitamin A).

Maintaining fluid balance

* The blood proteins albumin and globulin are important in maintaining fluid balance between the blood and the surrounding tissue space.
* Excessive fluid can build in the surrounding tissues because the counteracting force produced by the smaller amount of blood proteins is too weak to pull enough of the fluid back from the tissues into the bloodstream. As fluid builds up in the interstitial spaces, the tissues swell, resulting in medical condition of protein deficiency edema.

Contributing to acid-base balance

* Proteins play an important role in regulating acid-base balance and body pH
* Proteins are especially good buffers for the body because they have negative charges, which attract positively charged hydrogen ions. This allow them to accept and release hydrogen ions as needed to prevent harmful changes in pH.
* Acid is a byproduct of several chemicals, excess of which can disrupt normal body functioning and in severe cases result in coma or death. Proteins buffer this acid until it can be removed from the body by the kidneys or lungs.

Forming hormones, enzymes and neurotransmitters

* Amino acids are required for the synthesis of most hormones in the body. Some hormones, such as the thyroid hormones are made from only 1 amino acid, whereas others, such as insulin are composed of many amino acids.
* Many neurotransmitters (released by nerve endings), dopamine and norepinephrine (synthesized from the amino acid tyrosine),

and serotonin (synthesized from the amino acid are protein based). Amino acids are also required for the synthesis of enzymes.

Contribution to immune function

Antibodies are proteins and are key component of the immune system. Antibodies can bind to foreign proteins (called antigens) that attack the body and can avert their attack on target cells.

Forming glucose by gluconeogenesis

If carbohydrate intake is inadequate to maintain blood glucose levels, the liver (and kidneys, to a lesser extent) is forced to make glucose from the amino acids present in body tissues. This process is called gluconeogenesis.

Providing energy

- Under normal conditions, body cells use primarily fats and carbohydrates for energy. Proteins supply very little energy for healthy individuals. The amino acids which are not used for protein synthesis are broken down to provide energy, 1g of protein gives rise to 4.2kcal.
- When carbohydrate is not sufficient to meet the glucose needs for central nervous system, then some amount of protein will be converted to supply glucose or if it is above the caloric requirements then it will be metabolized into fat.

When we go through the details of proteins we come across many interesting facts about them and they are as:

Classification of proteins

On the basis of physical structure	Globular (Haemoglobin, Myglobin)
	Fibrous: two types:
	Elastic (Elastin)
	Non elastic (Collagen)
On the basis of nutritional importance	Complete (Egg)
	Partially complete (Zein)
	Incomplete (Glydin)
On the basis of chemical structure	Simple (Ovuglobulin and Prolamin)
	Conjugated (Metalloprotein and Lipoprotein)
	Derived (Peptones and Peptones)

- ★ Amino acids can be essential and non essential. For adults 8 and for children 9 and for chicks 10 amino acids are essential.
- ★ Conditionally essential amino acids are those in which the human body is unable to make sufficient amounts of that amino acids and hence all or a part of the daily needs of that amino acid must be provided by the diet.

Classification of amino acids:

Essential (Indispensable)	*Conditionally Essential*	*Non-essential (Dispensable)*
Methionine, Tryotophan, Threonine, Valine, Isoleucine, Lysin, Phenylalanne, Leucine.	Arginine, Cysteine, Glycine, Proline, Tyrosine	Proline, Hydroxyproline, Glutamic acid, Alanine

- ★ Arginine is required for chicks and histidine for children.
- ★ Four limiting amino acids are: methionine, lysine, threonine and tryptophan.
- ★ Asphaltamine, glutamic acid, tryptophan have high nitrogen content.
- ★ Methionine an essential amino acid form vitamin choline.
- ★ Choline can also form lecithin.
- ★ Tryptophan can form niacin with help of serotonin.
- ★ Albumin plays an important role in maintaining the osmotic pressure.
- ★ The nitrogen content of food-stuffs are converted into proteins by multiplying the factor 6.25.
- ★ Urea cycle involved three compounds- Arginine, Citruline and Ornithine.
- ★ Stomach secretes HCL.
- ★ HCl is secreted by parietal cells of stomach.
- ★ Pepsin is present as inactive pepsinogen and is activated by HCl and pepsin for protein digestion.
- ★ Typsinogen is present as inactive trypsinogen and activated by enterokinase.

- ★ Chemotrypsin is present as inactive chemotrypsinogen and activated by trypsin.
- ★ In pancreas, chyme activates duodenum to produce secretin (stimulates flow of pancreatic juice) and cholecystokinin (stimulates production of enzymes).
- ★ Cereals have 6-12g protein, pulses 17-25g, egg 12-14g and meat 15-20g protein.
- ★ Pulses are deficient in methionine.
- ★ Red gram is deficient in tryptophan.
- ★ Seasame seeds are rich in methionine.
- ★ Caseinogen of milk and milk proteins is rich in methionine.
- ★ The factor 14 in protein calculation in Kjeldhal method is due to the fact that 100ml of 1N hydrochloride neutralizes 14g of nitrogen.
- ★ Egg contain ovomucoid which inhibit trypsin activity.
- ★ Vegetable proteins are usually deficient in lysine, methionine or threonine making the quality of protein low.
- ★ Excess heating of vegetable protein causes loss of quality of lysine and methionine.
- ★ Siclar Williams introduced term Kawashiokar in 1931.
- ★ Jennifer introduced term PEM (Protein Energy Malnutrition) in 1959.

Features	***Kawashioskar***	***Marasmus***
Reason	Due to prolonged period of breast feeding resulting in poor nutrient intake.	Due to early abrupt weaning which prevent child from mother's milk and switched him/her on poor diet.
Odema	Present	Absent
Muscle wasting	Moderate	Severe
Hair changes	Present	Absent
Body weight standards (%)	About 60 %	50%
Moon Face	Present	Absent
Other features	Diarrhea, dermatitis, anorexia, grey hair, anaemia, angular stomatitis, cheilosis, fatty liver.	Diarrhea, anorexia, shrunken body, hungry victim

- ★ The edema of protein deficiency may be the result of the body's inability to regulate the protein hormone, particularly Anti-Diuretic Hormone (ADH) which plays a role in controlling water balance.
- ★ The RDA of proteins is dependent on requirement of maintenance of body, obligatory losses and growth.
- ★ There is always a balance between protein synthesis and protein degradation. If synthesis is more than breakdown then it is called as an anabolic state that builds lean tissues and if breakdown is more than synthesis then is known as catabolic state that burns lean tissues.

5
Carbohydrates in Human Nutrition

Introduction

- ★ Carbohydrates are nutrients made up of carbon, hydrogen and oxygen. Carbohydrates supply energy for body's functions that's why they are known as "A class of energy yielding substances'. They provide energy for the brain and ½ of energy for muscles and tissues. In form of dietary fibre, they contribute to bulk of stools and thus prevent constipation, colon cancer, diabetes, high blood pressure.

Classification of Carbohydrates

Class (DP)*	*Sub-Group*	*Components*
Sugars (1-2)	Monosaccharides	Glucose, galactose, fructose
	Disaccharides	Sucrose, lactose, trehalose
	Polyols	Sorbitol, mannitol
Oligosaccharides (3-9)	Malto-oligosaccharides	Maltodextrins
	Other oligosaccharides	Raffinose, stachyose, fructo-oligosaccharides
Polysaccharides (>9)	Starch	Amylose, amylopectin, modified starches
	Non-starch polysaccharides	Cellulose, hemicellulose, pectins, hydrocolloids

*** DP is degree of polymerization or number of sugar units.**

Carbohydrates

Key notes on carbohydrates

- ★ Reducing sugars are those that can reduce cupric ions to cuprous ions and has free aldehyde or ketone group. e.g. maltose and lactose.
- ★ Non reducing sugars can not convert cupric ions to cuprous ions. e.g. sucrose.
- ★ Amylose has 500-20,000 glucose units and alpha 1,4 glycosidic linkage.
- ★ Amylopectin has 500-1000 glucose units and has alpha 1,4 (main chain) and 1,6 linkage (branched chain).
- ★ Glycogen has 1,4 and 1,6 glycosidic linkage and can have 5000-10,000 glucose units.
- ★ Glycosidic linkage is linkage between two sugar units.
- ★ Glucose means "Sweet Wine" in Greek.
- ★ Fructose is also known as levulose.
- ★ On reduction, fructose form sorbitol and mannitol while, galactose form dulicitol.
- ★ Sucrose is also known as invert sugar, fructose as fruit sugar and lactose as milk sugar.
- ★ Breast milk have more oligosaccharides thus stimulate growth of bifidobacteria.
- ★ Gums lower cholesterol, have high water holding capacity and increase fecal bulk.
- ★ Raffinose is formed by one unit of glucose, one unit of fructose and one unit of galactose.
- ★ Stachyose is formed by one unit of glucose, one unit of fructose and two unit of galactose.
- ★ Polysaccharides can be classified on the basis of their digestibility:

  Digestible carbohydrates: Starch, dextrins, glycogen

  Partially digestible: Inulin, mannan and galactan.

  Undigestible: Cellulose, lignin, hemicellulose

★ The optimum pH for digestion by amylases is 6-6.9 and is activated by chloride ions.

★ In stomach, food get converts to chyme.

★ Salivary amylase works on cooked starch and pancreatic amylase works on raw starch.

★ Carbohydrates are transported by active transport involving ATPase pump in which and sodium and potassium play an important role.

★ The enzyme amylase is secreted by parotid, submaxiallry and sublingual glands of mouth.

★ Amylase is also known as ptylin and lipase as steapsin.

★ Pancreatic juice contains a carbohydrate splitting enzyme, pancreatic amylase (amylopsin) similar to salivary amylase.

★ Pancreatic amylase, an isoenzyme of salivary amylase, differs only in the optimum pH of action.

★ Enzymes require chloride ions for their actions (Ion Activated Enzymes).

★ In anaerobic glycolysis, lactic acid is produced.

★ Raffinose and stachyose, two oligosaccharides are known as gas forming carbohydrates as they are not completely digested by human body enzymes because of lack of enzyme α glactosidae (α- GAL).

★ Butyrate formed after fermentation of fibre in small intestine provide energy for colonocytes that protect against cancer.

★ Waxy corn has 2 percent amylase.

★ High amylase corn has 80 percent starch.

★ RDS is rapidly is digestible starch. *e.g.* Freshly cooked starchy foods like potato, processed foods.

★ Galactose in milk is essential for formation of mylin that transmit nerve messages.

★ SDS is slow digestible starch. *e.g.* Wholy or partly milled grain like pasta, legumes.

★ RS is resistant starch. *e.g.* Whole grain, unripe banana.

★ Starch has 30 percent amylase and 70 percent amylopectin.

- ★ Retrogradation is precipitation of amylase due to which starch solution get turbid.
- ★ Dextrin is formed by partial hydrolysis of starch.
- ★ Dextrimaltose is used in infant feeding.
- ★ Liver has 3-7 per cent stored glycogen.
- ★ Adult liver has 200g glycogen while muscle have around 350 g of stored glycogen.
- ★ Hemi cellulose is formed by pentoses, hexoses and uranic acid. It lower serum cholesterol and blood sugar.
- ★ Lignin, an undigestble carbohydrate provide resistance to cell wall from microbial degradation and has absorption power for cancer producing hydrocarbons.

Functions of Carbohydrates

- ★ They provide energy and serve as energy resource. They are the sources of metabolic fuels and energy reserve.
- ★ These are structural components of cell wall. In plants and of the exoskeleton of antropods (insect family).
- ★ Carbohydrates can also be used in making of nonessential amino acids, fats, components of DNA, RNA (deoxy and ribose sugar) and other structures and compounds in the body.
- ★ These are integral features of many proteins and lipids (glycolipids and glycoproteins). They are present in cell membrane.
- ★ Provide sweetness and humectant properties to food (to keep the product moist).
- ★ Galactose is component of nerve membrane.
- ★ Glucose is needed by brain 120g/day.
- ★ Add bulk to the diet (dietary fibers).

6

Fats in Human Nutrition

Characteristics Features

* Fats or oils are the most concentrated source of energy. In our daily diet 10–30 per cent of our energy needs is fulfilled by fat.
* Like carbohydrates, fats are organic compounds composed of carbon, hydrogen and oxygen. They differ from carbohydrates because they contain greater proportion of carbon than hydrogen and oxygen.
* Fats or oils contains 76 per cent carbon, 12 per cent oxygen and 12 per cent hydrogen. Fats vary in their structure due to the presence of different fatty acids.
* Fats are insoluble in water and soluble in organic solvents like ether, benzene or chloroform. Their cooking properties depend upon the type of fatty acids present in them.
* If the substance is liquid at 20°C, it is called oil, if it is solid at that temperature, it is known as a fat.
* The hardness, melting point and flavour of fat is related to the length of carbon chain and the level of saturation of the fatty acid.
* Saturated fatty acids are found in solid fats where as most of the oils contain unsaturated fatty acids. When the hydrogen is added to the oils, the unsaturated fatty acids can be converted to saturated ones, thus changing the physical and chemical properties of the oil.

★ Saturated fatty acids have all the carbon atoms in the chain saturated with hydrogen atoms. When double bond is present between two carbon atoms the fatty acid is termed as unsaturated fatty acid.

★ The degree of unsaturation varies according to the presence of double bonds. If there are more double bonds in a fat it is called polyunsaturated fatty acid.

★ Palmitic, stearic and butyric acids are saturated fatty acids. Oleic, linoleic, linolenic and arachidonic acids are unsaturated oils. They have one, two, three and four double bonds, respectively.

★ The unsaturated fatty acids are often described as essential fatty acids. Vegetable oils are rich in linolenic acid, fish oils and animal fats are rich in arachidonic acids.

★ Whenever we come across the term "Fats" we always think of their health consequences beyond their role as energy source to body. Fats are those macronutrients that are essential for growth and development and are responsible for nutrition-related chronic diseases in later life if taken in uncontrolled manner.

Some Important Functions of Fats and Fatty Acids

★ **Help body to use vitamins:** Vitamins A, D, E, and K are fat-soluble vitamins, which means that the fat in foods helps the intestines absorb these vitamins into the body.

★ **Brain development:** Fat provides the structural component not only to cell membranes in the brain, but also to myelin, the fatty insulating sheath that surrounds each nerve fiber, enabling it to carry messages faster.

★ **Energy:** Each gram of fat provides 9 calories of energy for the body, compared with 4 calories per gram of carbohydrates and proteins.

★ **Healthier skin:** One of the more obvious signs of fatty acid deficiency is dry, flaky skin. The layer of fat just beneath the skin acts as the body's own insulation to help regulate body temperature.

★ **Healthy cells:** Fats are a vital part of the membrane that surrounds each cell of the body. Without a healthy cell membrane, the rest of the cell couldn't function.

★ **Making hormones:** Fats are structural components of some of the most important substances in the body, including prostaglandins, hormone-like substances that regulate body functions.

★ **Palatability:** Besides being a nutritious energy source, fat adds to the appealing taste, texture and appearance of food. Fats carry flavor.

★ **Protective cushion for our organs:** Many of the vital organs, especially the kidneys, heart and intestines are cushioned by fat that helps protect them from injury and hold them in place.

Classification of Fats

Simple lipids- Neutral fat and Waxes

Compound lipids- Phospholipids *e.g.* Lecithin
Glycolipids *e.g.* Cerebrosides
Lypolipids *e.g.* Cholesterol

Derived lipids- Made up of simple and compound lipids.

Key Notes on Fats

★ Lipids are also divided into eight categories *viz.* Fatty acyls (oleic acid), Glycerolipids (triacylglycerol), Glycerophospholipids (phosphatidylcholine), Sphingolipids (sphingosine), Sterol lipids (cholesterol), Prenol lipids (farnesol), Saccharolipids [UDP-3-(3 hydroxytetradecanoyl- N-Acetylglucosamine], Polyketides (aflatoxin).

★ Gastric lipase initiates the secretion of cholecystokinin which causes secretion of bile from liver.

★ Gastric lipase acts only on short and medium chain fatty acids.

★ Gastrin stimulates secretion of gastric juice in gall bladder.

★ MCT (Medium Chain Triglycrides) can be given as it can be digested and absorbed in absence of bile salts in gall bladder diseases.

★ Beta oxidation is the oxidation of fatty acid in which α carbon are removed subsequently.

★ Iodine value is an index of unsaturation. More unsaturation more iodine value.

- ★ More peroxide value less stability of fats.
- ★ Choline, inositol and vitamin B12 are known as lipotropic substances.
- ★ Oleic acid has chemical formula $C_{18}H_{34}O_2$ (1 double bond), of linoleic acid $C_{18}H_{32}O_2$ (2 double bonds) and of linolenic acid $C_{18}H_{30}O_2$ (3 double bonds).
- ★ Linoleic and linolenic acid are known as essential fatty acids as they cannot be formed by our body.
- ★ Essential fatty acid requirement is 15-25/ day.
- ★ 100-200 g of fish 2-3 times a week prevents CHD.
- ★ The examples of MUFA are: oleic & Euracid acid found in olive oil, canola oil, almond oil and ground nut oil.
- ★ The examples of PUFA are: Linolenic acid, ecosapentanoic acid & necosa penta noic acid (w3) & linoleic & arachidonic acid (w6)
- ★ Egg has cholesterol content of 250mg cholesterol (permitted is 300mg/day).
- ★ Sterols are present in animals as cholesterol and in plants as ergosterol.
- ★ Trans-Fatty Acids (TFA) are produced when vegetable oils are hydrogenated. Dietary trans-fatty acids are derived from dairy and meat products and from products made from hydrogenated fats. The main modifiable source of TFA includes partially hydrogenated vegetable oils (PHVO) such as vanaspati, bakery fats and margarines.
- ★ A healthy adult man of 70kgs has 10-15kg of adipose tissues and woman of 60kgs has 10-20kgs of adipose tissues.
- ★ Brown adipose tissues are present in new born animal and human infants and known as mini- furnance.
- ★ Swimmers are more fat (fat floats) while wrestlers/gymnasts want less fat. Endurance runners don't want massive amounts of muscle but power lifters do.
- ★ Bilirubin is formed in macrophages of the reticuloendothelial system. The initial substrate is predominantly hemaglobin.
- ★ The products of lipid digestion coalesce with bile acids into mixed micelles.

★ Phospholipase A_2, pancreatic lipase and cholesterol esterase are key enzymes involved in fat digestion.

Dietary Sources of Cholesterol

Type of Fat	*Main Source*	*Effect on Cholesterol Levels*
Monounsaturated	Olives, olive oil, canola oil, peanut oil, cashews, almonds, peanuts and most other nuts, avocados	Lowers LDL, Raises HDL
Polyunsaturated	Corn, soybean, safflower and cotton seed oil, fish	Lowers LDL, Raises HDL
Saturated	Whole milk, butter, cheese and ice cream; red meat; chocolate; coconuts, coconut milk, coconut oil, egg yolks, chicken skin	Raises both LDL and HDL
Trans	Most margarines; vegetable shortening; partially hydrogenated vegetable oil; deep-fried chips; many fast foods; most commercial baked goods	Raises LDL

★ Fatty acids helps in production of hormone-like substances called Eicosanoids which are responsible for regulating blood pressure, blood viscosity, vasoconstriction, immune and inflammatory responses.

★ Deficiency of fatty acids causes phrynoderma in which skin become rough and thick horny papules of pin head size erupt.

★ On body's demand pyruvic acid can be converted into triglycerides.

Recommended Dietary Allowance (RDA)

★ Since human requirement of fat is not known, no specific recommendation is yet made regarding the amount of fat in the diet. However, in order to meet the essential fatty acid need, diet should contain at least 35 g of vegetable oils.

Food Sources

★ Vegetable oils, like groundnut oil, coconut oil, mustard oil. Hydrogenated oils butter, ghee and curd are the common sources of fats. Butter and ghee are animal fats extracted from milk.

Table 6.2: Fatty Acid Composition of some Edible Oils and Fat

Fat/Oil	Palmitic	Stearic	Arachidic	Behenic	Lignoceric Saturates	Total	Palmitoleic	Oleic Mono- saturates	Total	Linoleic	a-Linolenic Poly Unsaturates	Total
16:0	18:0	20:0	22:0	24:0		16:1	18:1		18:2	18:3		
Coconut oil	7.8	2.3	–	–	–	89.5*	–	7.8	7.8	2.0	–	2.0
Corn oil	10.7	1.7	–	–	–	12.7	–	29.6	29.6	57.4	–	57.4
Groundnut oil	12.6	1.7	4.2	2.1	0.3	20.9	1.4	47.9	49.3	29.9	–	29.9
Mustard oil	2.9	0.9	6.9	–	–	10.7	0.6	8.9	56.0**	18.1	14.5	32.6
Olive oil	12.5	2.3	–	–	–	14.8	–	74.5	74.5	10.0	–	10.0
Palm oil	42.0	4.3	–	–	–	46.3	–	43.7	43.7	10.0	–	10.0
Palmolein	42.3	3.5	–	–	–	47.7	0.4	41.0	41.4	10.3	0.3	10.6
Rice bran oil	19.5	2.6	–	–	–	22.1	–	41.0	41.0	34.3	1.4	35.7
Safflower oil	7.8	2.1	0.8	–	–	10.7	–	16.7	16.7	73.5	–	73.5
Gingelly oil	9.7	4.0	–	–	–	13.7	0.1	41.2	41.3	44.5	–	44.5
Soyabean oil	9.8	2.4	0.9	–	–	13.1	–	28.9	28.9	50.7	6.5	57.2
Sunflower oil	5.6	2.2	0.9	–	0.4	9.1	–	25.1	25.1	66.2	–	66.2
Butter	29.4	13.5	–	–	–	69.4*	–	28.0	28.0	2.5	–	2.5
Lard 26.2	20.0	–	–	–	46.2*	–	45.2	45.2	11.0	–	11.0	
Tallow	27.9	21.3	–	–	–	54.9*	–	40.9	40.9	4.2	–	4.2

*Includes lower chain fatty acid

** Includes 46.5 per cent of erucic acid (22:1).

Source: Nutritive value of Indian foods, C. Gopalan et al., NIN, ICMR.

7

Minerals in Human Nutrition

Introduction

- ★ Mineral elements form an important group of nutrients necessary for the growth and upkeep of the body. About 4–6 per cent of body weight is made up of mineral elements.
- ★ There are 24 minerals in the body. Mineral substances are important in maintaining good health because they are part of our tissues, skeletal structure and body fluid. Hormones, enzymes and vitamins contain minerals.
- ★ Vital organs like heart and nervous tissues cannot function normally in the absence of minerals.
- ★ Minerals are inorganic substances that are present in the body tissues like phosphomin, iron, sulphur, zinc or copper or soluble minerals in the body fluids like sodium chloride or mineral components of the bones, teeth and hormonal secretions.

Functions

- ★ Sodium and potassium regulate the water balance. The minerals, calcium and iron are normally included in nutritional planning. The largest concentrate of minerals is found in bones and teeth. Few milligrams of mineral elements decide the skeletal growth or the mental ability of persons.

★ Optimum intake of minerals and favourable environment for their utilization are equally important to maintain mineral balance.

★ During the growth period and also during pregnancy, lactation and old age, certain minerals are very essential to maintain health and growth.

★ Minerals do not act singly in their regulation of body processes but work with the help of other and organic compounds. About 850–950g of calcium is present in a healthy man.

★ Minerals are inorganic substances needed by the body for many different functions. Some minerals are needed in very tiny amounts, thus they are called as micronutrients.

★ Minerals have different functions and they are as:

- *Structural*: They form an integral part of structures such as the bones/skeleton, blood etc.
- *Catalytic*: Certain minerals are required as constituents of enzymes, coenzymes in various metabolic pathways.
- *Cellular*: Some are necessary for membrane stability, as well as, inter and intra cellular transport mechanisms.
- *Others*: They play an important role in muscle contraction, nerve transmission etc.

Classification

★ They are generally classified in three categories:

- Major minerals- required in large amounts at least 100mg/day *e.g.* Phosphorus, sodium, chlorine, potassium.
- Minor minerals- required in small quantities less than a few mg/day *e.g.* Sulphur, magnesium, iron.
- Trace elements- required in micrograms *e.g.* Iodine, zinc, fluorine.

Key notes on Minerals

★ Anaemia is a condition in which the body does not have enough healthy red blood cells.

- ★ Transferrins are iron-binding blood plasma glycoproteins that control the level of free iron in biological fluids
- ★ Hypochromic anaemia is a type of anaemia in which RBCs become small and pale also known as microcytic anaemia (deficiency of iron).
- ★ Orthochromic anaemia is characterized by large, irregularly shaped RBCs. Also known as macrocytic (deficiency of iron, vitamin B_{12}).
- ★ Dimorphic anaemia is dual population of RBCs in the blood with both microcytic hypochromic and normochromic macrocytic RBCs (due to deficiency of iron, Vitamin B_{12} or folic acid or both).
- ★ Pernicious anemia is due to deficiency of vitamin B_{12} due to failure in intrinsic factor production by stomach cells.
- ★ Sideroblastic anaemia is a disease in which the bone marrow produces ringed sideroblasts rather than healthy red blood cells.
- ★ According to national nutritional anaemia prophylaxis programme (NNAPP) for an anemic pregnant woman 60mg of elemental iron and 500µg of folic acid should be provided and for children the dose should be 20mg iron and 100µg of folic acid.
- ★ Kolynchia is a symptom of anaemia in which nails become spoon shaped.
- ★ Egg has phospholipid inhibitors due to which iron is poorly absorbed.
- ★ Thalassaemia is defect in synthesis of a part of polypeptide chain of hemoglobin.
- ★ Iron is transported in human body as ferritin and transported as haemosiderin.
- ★ When RBCs disintegrate non protein haemoglobin is splitted into haematin and then to billirubin.
- ★ WHO's Hemoglobin thresholds used to define anemia:

Age or Gender Group	***Hb Threshold (g/dl)***
Children (0.5–5.0 yrs)	11.0
Children (5–12 yrs)	11.5
Teens (12–15 yrs)	12.0
Women, non-pregnant (>15yrs)	12.0
Women, pregnant	11.0
Men (>15yrs)	13.0

- ★ Iron from fortified sugar would be expected to be well absorbed if consumed with citrus drinks but poorly absorbed from coffee and tea owing to phenolic compounds or, if added to cereal products, owing to phytate.
- ★ Inorganic iron binds to phytates, tannins and phosphates in plant foods and these may have an inhibitory effect on iron absorption.
- ★ Each gram of haemoglobin contains 3.34mg of iron.
- ★ Iron is absorbed in ferrous form.
- ★ In National Nutrition Anaemia Prophyloxis Programme recomended supplementation of iron & folic acid for women's 60 mg & 500 μg respectively for pre-school children this dosage is 20 mg & 100 μg, respectively.
- ★ Iron in form of ferric ammonium citrate is one of the best tolerated iron supplements.
- ★ In fortification of iron, ferrous is microencapsulated with zein to prevent its oxidation.
- ★ Daily consumption of 10 g salt (25ppm) of potassium iodate provides 150 μg of iodine.
- ★ Daily consumption of 10g of iodized salt (25ppm of potassium iodate) provide only about 150 μg of iodine.
- ★ Goitrogens can block uptake of iodine by thyroid gland.
- ★ NIDDCP is National Iodine Deficiency Disorder Control Programme.
- ★ Potassium iodate is more stable than potassium iodide.
- ★ Thyroid peroxidase (TPO) is important in the process of attaching iodine to the tyrosine for the production of T_4 and T_3.
- ★ Excess phosphates, phytic acid, fatty acid, fibre, oxalic acid inhibit calcium absorption while protein, lactose, normal pH not alkaline, citric acids help in its absorption.
- ★ During rapid growth, calcium to phosphorus ratio needed in diet is 1:1 which decreases to 1:2 with attainment of adulthood and in female in case of pregnancy and lactation this ratio should be 1:1.

- ★ Calcium is required for vascular contraction and vasodilation, muscle function, nerve transmission, intracellular signaling and hormonal secretion.
- ★ Parathyroid regulates calcium level in blood.
- ★ Potassium ions increases the relaxation of heart muscles which is antagonized by calcium ions.
- ★ RBCs contain potassium ions.
- ★ Insulin contains zinc in form of enzyme carbonic anhydrase.
- ★ Egg is a source of biologically active zinc but Fe is bound to conalbumin and is poorly absorbed
- ★ Zinc is required for protein, DNA and RNA synthesis.
- ★ Zinc is essential for healthy appetite, perception of taste and vision in night.
- ★ The vegetarian diet is low in zinc. Animal foods-meat, eggs, oysters and other sea food are rich in zinc.
- ★ Magnesium acts as activator of several enzymes *i.e.,* alkaline phosphates and phosphorylating enzymes.
- ★ Manganese deficiency can case ataxia, impaired growth, depressed reproductive function while, toxicity causes blurred speech and spastic gait.
- ★ Sulphur can exist in the reduced form in cysteine and oxidized in cystine form.
- ★ Sodium is extracellular and potassium is intracellular ion.
- ★ Fluorosis occurs when fluoride content of water is more than 3-5 ppm.
- ★ Dental fluorosis is a change in the appearance of the tooth's enamel.
- ★ When too much water is lost from body, to maintain balance hypothalamus releases ADH due to which kidneys reabsorb more water and retain sodium.
- ★ Menkey Kincky Hair Syndrome is caused by copper deficiency.
- ★ Toxicity of copper is called Wilson's disease and is characterized by green rings in eyes.

* Selenium is constituent of enzyme glutathione peroxidase which is present in cytosol and prevent the formation of free radical.
* Lithium can block the release of thyroid hormones from the thyroid gland.
* Sodium is required for absorption of monosaccharide and amino acids.
* Because of its capacity to "mimic" calcium, lead is stored in the bones and becomes a stable bone component, particularly in the case of insufficient calcium intake. This lead deposit can mobilize and return into the blood stream resulting into anemia.
* Minamata disease is associated with mercury poisoning.
* Chromium form complexes with nicotinic acid and amino acids to form GTF. Being a part of the 'Glucose Tolerance Factor' chromium has an impact on glucose tolerance. Chromium improves insulin function and plays a role in glucose metabolism.

Recommended Dietary Allowance (RDA)

* The FAO/WHO expert group (1984) has recommended 400–500 mg/day calcium for an adult man and woman. (During pregnancy and lactation the requirement is one g/day).
* Phosphorus : Requirement is one gm/day.
* Iron: Requirement is calculated on the basis of its loss and absorption. For adult man weighing 55 kg weight it is 24 mg/day and for women of 45 kg, 32 mg/day. For infants till one year 1 mg/kg body weight is required.
* Sodium: For an adult 1–2 g of sodium and for children 350 mg is recommended. For infants 150–400 mg is enough.
* Iodine: Requirement of iodine for an adult is about 0.15 to 0.2 mg/day and for infants and children 0.05 to 0.1 mg/day is enough.
* Copper: The requirement of copper is about 2 mg/day for adult. During pregnancy and lactation the copper requirement is increased to 3 mg/day.

- Zinc: Infants require about 3 to 5 mg/day and for children it is 10 to 15 mg/day. During pregnancy and lactation 20 to 25 mg of zinc is required.

Sources of Minerals

- Milk and milk products are the best sources of calcium. Ragi is the cheapest source of calcium therefore it is known as poor man's milk.
- Vegetables like peas, beans, potatoes, pulses and small fish and dried fruits are rich in calcium.
- Animal foods like meat, fish, poultry and eggs are excellent sources of phosphorus.
- Milk and milk products are good sources of phosphorus.
- Liver, kidney, heart, lean meat, egg yolk and shellfish are the best sources of iron.
- Animal foods like meat, meat extractions, fish poultry, milk, milk products, nuts, bakery items, salted biscuits, egg white, dried fruits, roots like beetroot, carrot, radish and leafy vegetables like spinach are excellent sources of sodium.
- The best sources of iodine are sea foods, common salt and vegetables. Cereals and legumes are poor sources of iodine.
- The iodine content of food depends on the iodine content of the soil.
- Millets, bajra, raw rice, whole gram, cow gram dried peas, red gramdal, vegetables like drumstick are good sources of copper.
- Liver is the richest source of copper.
- Seafood, meat, poultry and eggs are good sources of zinc, cereals, legumes and nuts contain some amount of zinc.
- On milling 80% of zinc is lost. Water containing 1 to 2 ppm fluorine prevents carries and so it is considered as the body requirement for fluorine.

Factors Influence Mineral Bioavailability

- The bioavailability of minerals can be greatly influenced by the amount of minerals consumed because many minerals have similar molecular weights and charges (valences).

- ★ For example, magnesium, calcium, iron and copper can each exist in the 2+ valence state. These minerals can compete with each other for absorption, thereby affecting each other's bioavailability.
- ★ As an example, an excess of zinc in the diet can decrease the absorption and metabolism of the mineral copper.
- ★ Mineral bioavailability also is strongly affected by the presence of non-mineral substance in the diet which interfere with absorption of mineral. These substances are fibers, oxalates, polyphenolic compounds etc.
    - The components of fiber, especially phytic acid (phytate) in wheat grain, can limit the absorption of some minerals by chemically binding to them and preventing their release during digestion.
    - Intake of dietary fiber greatly above the adequate level of 25 to 38 g/day can adversely affect mineral status and availability. However, if grains are leavened with yeast, enzymes produced by the yeast can break some of the chemical bonds between phytic acid and minerals.
    - Oxalic acid (oxalate) found in leafy green plants binds minerals and makes them less bioavailable.
    - Spinach contains good amount of calcium, but only about 5 per cent of it can be absorbed because of the high concentration of oxalic acid in spinach.
    - Polyphenols also can lower the bioavailability of minerals, especially iron and calcium.
    - Mineral bioavailability can be enhanced by some vitamins like vitamin C, which can improve absorption of iron when both are consumed in the same meal.
    - HCl provides an electron to ferric iron (Fe^{3+}) to yield ferrous iron (Fe^{2+}), which is better absorbed than ferric iron. Use of antacids and reduced stomach acid production (which is common in old age) can hinder mineral bioavailability.

8

Vitamins in Human Nutrition

Introduction: An Overview

- ★ Vitamins may be regarded as organic compounds required in the diet in small amounts to perform specific biological functions for normal maintenance of optimum growth and health of the organism.
- ★ Vitamins can not be synthesized by our body. We must obtain them from all food we eat or by supplements.

Classification of Vitamins

Fat Soluble	*Water Soluble*		
	Non B-complex	*B-complex*	
		Energy Releasing	*Hematopoietic*
Vitamin A	Vitamin C	Thiamin (B_1)	Folic acid (B_9)
Vitamin D		Riboflavin (B_2)	Cynocobalamin (B_{12})
Vitamin E		Niacin (B_3)	
Vitamin K		Pantothenic Acid (B_5)	
		Pyridoxine (B_6)	
		Biotin (B_7)	

Sources of Vitamins

- Vitamin A is present in animal foods only. Liver is the richest source of Vitamin A. Other sources include butter, ghee, milk, curd and egg yolk.
- Refined oils and vanaspati are good sources of Vitamin A if these are fortified with vitamin A to the extent of 750 ug per 100 g. Vitamin A is not present in vegetable foods but these foods contain the pigment, β-carotene, which is precursor of vitamin A and is therefore also known as 'Pro-vitamin A'.
- Leafy vegetables such as spinach, amaranth, coriander, drumstick leaves as well as ripe fruits such as mango, papaya and yellow pumpkin are good sources of β-carotene.
- Foods are not good sources of Vitamin D. It is found in small quantities in liver, egg yolk, milk and milk fat (butter and ghee), obtained from animals fed on pastures exposed to sunlight.
- Almost all food contains some amount of thiamine, except sugars, fats and oils. Plant sources include pulses, nuts, oilseeds and whole grain cereals.
- Parboiled rice and fresh peas are good sources of thiamine. Green leafy vegetables and animal foods such as milk, eggs, fish and meat are fair sources.
- The main sources of thiamine in Indian diet are cereals but refining of cereals reduces thiamine content with most of the thiamine being lost in the polishing.
- Parboiling of rice helps to conserve thiamine. If cooking water is not discarded, a minimum mineral quantity of Thiamine is lost.
- Milk is the richest source of riboflavin. There is no riboflavin in butter and ghee because the vitamin is water-soluble and remains in the water extract during the removal of butter from milk or curd.
- Liver and kidney of animals and birds are also good sources of riboflavin.
- Meat is a rich source of niacin, the higher concentrations found in organ meat such as liver groundnuts are the best source of niacin.

- ★ Cereals are the major source of niacin in the Indian diet, unrefined and parboiled cereals retain more niacin than refined ones.
- ★ Amla is one of the richest source of ascorbic acid. Citrus fruits such as oranges, sweet lime, grape fruit are also excellent sources of ascorbic acid.
- ★ Guava, cashew apple and drumsticks have a high ascorbic acid content. Sprouted pulses such as bengal gram and green gram are good sources and have proved to be of value during drought and famine conditions when other sources of ascorbic acids, such as fresh fruits and vegetables are not available.
- ★ The vitamin requirements are small but they perform specific and vital functions essential for maintaining health.
- ★ Each vitamin has its own unique function required for the body for maintaining the growth.

Key notes on Vitamins

- ★ There are about 24 minerals and 13 vitamins.
- ★ Fat soluble vitamins are stored in body hence can cause toxicity. While, water soluble vitamins cannot be stored in body. They are excreted in urine and have no toxicity.
- ★ Beta carotene is the precursor of Vitamin A.
- ★ Vegetables sources contain the provitamin A- carotenes. Yellow and dark green vegetables and fruits are good sources of carotenes. *E.g.* carrots, spinach, amaranthus, pumpkins, mango, papaya etc.
- ★ One microgram of carotene is equal to 6 μg of beta-carotene.
- ★ 0.6 μg carotene is equal to one IU of vitamin A.
- ★ 0.3 μg of retinol is equal to 1 IU of vitamin A.
- ★ RBP is Retinol Binding Protein.
- ★ Cones in the retina are responsible for vision under bright lights. It translates objects to color vision.
- ★ Rods in the retina are responsible for vision in dim lights, translate objects to black and white vision.
- ★ Retinoic acid cannot be converted to retinol hence cannot help in visual cycle.

★ The Vitamin A deficiency may lead to ocular defects, the stages of which are:

- *Night Blindness*-Night blindness is the earliest symptom of Vitamin A deficiency in pre-school children.
- *Conjunctival xerosis* (XIA) – After night blindness, if deficiency continues, the condition further deteriorates and the conjunctiva becomes. dry and non wettable due to reduction of goblet cell mucus. It appears muddy and wrinkled instead of shiny and smooth.
- *Bitot's spot* (XI B) – This is further the advance stage. These are triangular, pearly white or yellowish, foamy spots on the bulbar conjunctiva on either side of the cornea.
- *Corneal xerosis* (X2) – In this stage cornea is affected which appear dull, dry and non wettable and eventually opaque.
- *Corneal ulceration* (X3A) – If corneal xerosis not treated promptly, it leads to ulceration of cornea.
- *Keratomalacia* (X3B) – This stage marks liquefaction of cornea. Cornea may become soft and may burst open. If the eye collapses vision is lost.
- *Corneal scar* (XS) – Corneal ulcer on healing leaves a white scar which may vary in size depending upon the size of the ulcer. If the scar blocks the papillary region, it affects normal vision.

★ The vitamin D has two forms: 25 (OH) D & 1, 25 (OH_2)D *i.e.*25 hydroxy cholecaliferol is formed in liver while 1,25 $(OH)_2$ D *i.e.* 1,25 dihydroxy cholecalciferol is released by kidneys and is also known as erythropoietin, a hormone.

★ 7-dehydrocholesterol present in skin form vitamin D in presence of sunlight.

★ 1 IU of vitamin D is equal to 0.025 mcg of cholecalciferol.

★ Vitamin D_2 is formed from ergosterol and is present on plants.

★ Vitamin D_3 is found in animals.

★ Vitamin D_3 is synthesized in the skin by ultra violet rays of sunlight

★ Milk is not a good source of Vitamin D.

* During exposure to solar ultraviolet B (UVB) radiation, 7- dehydro-cholesterol in the skin is converted to pre- vitamin D_3, which is immediately converted to vitamin D_3 in a heat-dependent process.
* Excessive exposure to sunlight degrades previtamin D_3 and Vitamin D_3 into inactive photoproducts.
* Rickets described by Glisson in 1650 in England among children on smoky areas. Disease is recognized after 6 months of deficiency.
* Vitamin E protects RBC from hemolysis by oxidizing agents.
* As an antioxidant, Vitamin E acts as a peroxyl radical scavenger, preventing the propagation of free radicals in tissues, by reacting with them to form a tocopheryl radical which will then be oxidized by a hydrogen donor (such as Vitamin C) and thus return to its reduced state. As it is fat-soluble, it is incorporated into cell membranes, which protects them from oxidative damage.
* A high intake of Vitamin E interferes with the functions of other fat soluble vitamins such as Vitamin K absorption and Vitamin D in terms of bone mineralization while in case of Vitamin A deficiency, it lowers the rate of Vitamin A depletion from the liver. However, large doses of Vitamin E inhibit β-carotene absorption or conversion to retinol in the intestine.
* Bile is essential for Vitamin K absorption.
* Vitamin K_1 is known as phylloquinone, K_2 as menaquinone and K_3 as menadione.
* Coumarin derivatives such as Warfarin and Dicumarol interfere with recycling of vitamin K and thereby lead to Vitamin K deficiency.
* An inter-relationship exists between Vitamin D and K based on their relationship to the mineral calcium. Vitamin D affects calcium metabolism and Vitamin K-dependent proteins bind calcium. The two sites of action of Vitamin D are bone and kidney tissues, where Vitamin K-dependent calcium-binding proteins have been identified which regulate the production of crucial enzymes.
* Thiamine is known as appetite Vitamin.
* A small part of thiamine is converted to TTP (Thiamine Triphosphate) and a large part is converted into TPP in (Thiamine Pyrophosphate) in liver.

- ★ Thermal destruction of thiamine results in meaty flavor in cooked foods.
- ★ Thiamine is sensitive to sulphur dioxide and sulphites.
- ★ About 60 mg of tryptophan is equal to 1mg of niacin.
- ★ Riboflavin is required for metabolism of amino acids.
- ★ Riboflavin and Vitamin B_{12} are present more in animal foods.
- ★ Angular stomatitis is infection of skin at the angles of mouth due to deficiency of riboflavin, thiamin and pyridoxine.
- ★ Galactoflavin is an antagonist of riboflavin.
- ★ Baking soda destroys riboflavin.
- ★ Flavin Mononucleotide (FMN) and flavin adenine dinucleotide (FAD) are two forms of riboflevin.
- ★ Glossitis is swollen reddish or bluish tongue with ulcers due to deficiency of niacin and cyanocobalamin.
- ★ Niacin deficiency is more in those areas where staple food is maize as it has high leucine content which interfere with niacin metabolism.
- ★ Anti tuberculosis drug isoniazid causes pellagra by forming inactive complex with niacin.
- ★ Niacin is chemically synonymous with nicotinic acid although the term is also used for its amide (nicotinamide).
- ★ It is the amide form that exists within the redox-active co-enzymes, nicotinamide adenine dinucleotide (NAD) and its phosphate (NADP).
- ★ Excess of leucine causes amino acid imbalance and decrease conversion of tryptophan to niacin.
- ★ Dietary factors such as deficiency of biotin, pantothenic acid can change colour of skin.
- ★ Burning feet problem is due to deficiency of pantothenic acid.
- ★ Vitamin B_5 *i.e.* pantothenic acid can be synthesized by intestinal bacteria.
- ★ Active form of pyridoxine is pyridoxal phosphate. Pyridoxal phosphate act as co enzymes in metabolism of amino acids.

- Biotin *i.e.* Vitamin H is also known for Anti egg white injury factor.
- Biotin is essential for the synthesis of malonyl CoA from acetyl COA and oxaloacetic acid from pyruvic acid.
- One mol avidin can bind 3mol of biotin.
- SIDS is Sudden Infant Death Syndrome also known as Cot death.
- Vitamin B_{12} is present only in animal foods. Lacto vegetarians get B_{12} through fermented milk. Vegans may have to take effort in meeting the needs of vitamin B_{12} by supplementation.
- Vitamin B_{12} act as coenzyme in the synthesis of methionine.
- Groundnuts are deficient in Vitamin A, Vitamin C and in lysin.
- Vitamin C is required for Metabolism of tyrosine
- Ascorbic acid is required for conversion of phenylalanine to tyrosine, epinephrine synthesis and oxidation of tyrosine and wound healing.
- Bleeding gums and loose teeth are symtoms of Vitamin C deficiency.
- Vitamin C is a powerful antioxidant because it can donate a hydrogen atom and form a relatively stable ascorbyl free radical (i.e. L-ascorbate).
- Fish Rohu is rich in Vitamin C.
- Folic acid -B_9 or Vitamin M Enhances brain health, maturation of blood cells.
- Follicular hyperkeratosis or phrynoderma is roughness of skin with horny plugs from follicular orphis.
- Xeroderma is characterized by dried skin with rough texture.
- Flaky paint dermatosis is hyper pigmented flaky skin.
- Creezing paving skin is a condition when skin appears as liquor painted and has formation of islands.
- Pellagra is known as 3D disease.
- Dry chopped appearance of lips with superficial ulcers is known as cheilosis.
- For production of RBCs following nutrients are required: Iron, folic acid, Vitamin B_{12}, protein, pyridoxine, ascorbic acid, copper and Vitamin E.

9

Dietetics: An Introduction

★ Dietetics is a science that deals with the adequacy of diets during normal life cycle and modifications required during diseased conditions. The aim of diet planning is:

- To maintain good nutritional status.
- To correct deficiencies.
- To afford rest to the particular body organ which is affected.
- To maintain the body's ability to metabolize the nutrients.
- To bring about the changes in body weight whenever necessary.

Food exchange lists

- They are the basis of meal planning they are groups of measured foods of the same calarific value & similar protein, fat & carbohydrate. All food exchange lists make aspecific contribution to a good diet. Exchange lists are based on principles of good nutrition that apply to everyone though extreme helpful for diabetics (Table 1 & Table 2).

Table 11.1: Food Exchange Lists 100 kcal. exchange list capacity of one standard katori is 130 ml.

Cereal exchange – 1.5 – 3.5 g protein	
Idli (big)	1
(medium)	11/3
Dosa (small)	1
(big)	1/2
Phulka	2
Chapathi	1
Puri	11/2
Rava Idli	1 Veg.
Sandwich	3/4
Bread Toast (medium)	11/2
Bread Pakoda	3
Plain rice	3/4 K
Upma	1/2 K
Veg. Noodles	1 K
Coconut Rice	1/2 K
Pongal	1/2 K
Boiled Wheat Rava	1 K
Sweet Pongal	2 Tb Sp
Pariyaram	1¼ pieces
Verrmicelli payasam	2
Kesari	11/2 Tbsp
Rice flakes upma	1/2 K
Naan	2/3
Cheese Sandwich	1/3
Ragi puttu	3/4 K
Ragi adai	3/4

Pulao	1/2 K
Bise bela bath	1/2 K
Tamarind rice	1/2 K
Curd rice	1/2 K
Idiappam	1
Pulse Exchange 3-5 g protein	
Sambar	11/2 K
Rasam	21/2 K
Thickdal	1/2 K
Thin dal	1 K
Channa masala	1/2 K
Dry peas sundal	3/4 K
Roasted bengal gram chutney (without coconut)	1/2 K
Sprouted moong salad	1 heaped katori
Pesarattu	3/4
Baked Masala Vada (with negligible amount of fat)	2 nos
Adai	1
Vada	1
Keerai	3
Vada	4
Bajji	2
Bonda	1
Meat exchange 5 g of protein	
Egg omelette	1
Scrambled egg	One egg
Fish Kolambu	2/ 3K
Fish fry	1 small piece

Boiled egg with gravy	1/3 to ½ serving
Meat curry	1 serving
Egg Custard	1/2 K
Milk exchange – 4.5 g protein	
Milk	150 ml – 1 tea cup
Curd	150 ml – 1 full katori
Cheese	11/3 cube 30g
Paneer	40 g
Butter milk	1 glass – (350 ml)
Badam Milk Shake	1/4 glass
Banana Milk Shake	1/3 glass
Milk kheer	1/2 K
Carrot Kheer	3/4 K
Vegetable A exchange	
Curry without coconut and gravy (with simple seasoning) Amaranth curry	1 heaped Katori
Plain tomato soup	2 K (1 soup bowl)
Soup with white sauce	1 K (½ soup bowl)
Mint chutney	1/2 K
Onion Chutney	1/2 K
Vegetable B exchange	
Roots and tubers curry	½ K
Cutlet	½
Fruit exchange	
Apple	1 medium
Banana	1
Custard apple	1
Pine apple	3 slices
Orange	2½

Sapota	2 (small)
Fruit salad (no sugar or dressing)	¾ K
Guava	1 (Big)

Source: Srilakshmi, B. Dietetics.2008. Menu Planning. New Age international Ltd Publishers

Nutrient intake is recommended to an individual on the basis of age, sex, weight, height, physical activity and physiological needs of the patient. It is always better to prescribe and formulate a suitable individual diet. Nutrition recommendations for patient are based on nutrition assessment, desired treatment outcomes and modification of usual food intake. For planning a therapeutic diet for a patient, first we should have the detailed knowledge about the body organ functioning and disease.

10
Human Body Organs & their Disorder

★ Out of 78 body organs in human body, 10-15 organs are crucial for its normal functioning. Any small defect in their functioning can lead to serious health hazards. Lets review some amazing facts about our body functioning, vital body organs and the disorders associated.

Body's immune system

★ Immunoglobulins have 5 classes: IgG, IgA, IgM, IgE,and IgD.

★ B-lymphocytes are produced and secreted in bone marrow.

★ The first line of defense (immunity) is provided by skin, mucus and their secretions.

★ Immunity types:

a. Adaptive immunity

Natural: Passive (Maternal), Active (infection)

Artificial: Passive (Antibody transfer), Active (Immunization)

b. Innnate immunity

★ Normal flora of human body are the microbes, mostly bacteria, that live in and on the body with, usually, no harmful effects to us. We have about 10^{13} cells in our bodies and 10^{14} bacteria, most of which live in the large intestine. There are 10^3–10^4 microbes per cm^2

on the skin (*Staphylococcus aureus, Staph. epidermidis*, Streptococci, *Candida, etc.*).

- ★ Cilia of respiratory tract, traps bacteria in mucus.
- ★ Prostaglandins help to produce fevers, sensitize pain receptors, cause contractions of the uterus, stimulate digestive tract motility, control nerve impulse, regulate blood pressure, promote blood clotting and cause inflammation.
- ★ Drugs like aspirin are effective in fever because they inhibit prostaglandin synthesis.
- ★ NSAID's is Non Steroid Anti-inflammatory Drugs which inhibit prostaglandins which in turn increases tissue exposure to acid and pepsin.
- ★ Aspirin blocks an enzyme called cyclo-oxygenase which is involved with the ring closure and addition of oxygen to arachidonic acid converting to prostaglandins. Asprin has the effect of blocking the channel in the enzyme and arachidonic cannot enter the active site of the enzyme. By inhibiting or blocking this enzyme, the synthesis of prostaglandins is blocked, which in turn relives some of the effects of pain and fever.

Gastrointestinal Tract

- ★ Diseases of oesophagus, stomach and deudenum are esophagitis, hernia, peptic ulcer.
- ★ Diseases of small intestine and colon are diarrhea, constipation, granulomatous colitis, diverticulosis, ulcerative colitis.
- ★ Melena is the passage of shiny sticky pitch black stools.
- ★ Hematemesis is presence of blood in vomiting.
- ★ *H. pylori* - a spiral, urease producing flagellated bacterium which lives between the mucus gel and mucosa.
- ★ Peptic ulcer is presence of ulcers in stomach walls.
- ★ Celiac disease is also known as gluten intolerance.
- ★ Peptic ulcer can also be caused by *Helicobacter pylori.*
- ★ Constipation can be atonic, spastic and obstructive.

★ Increased acid secretion can lead to gastritis.

★ Antacids neutralizes the presence of acids in stomach.

★ Inflammatory bowel diseases is of two types: Crohn's disease and Ulcerative colitis.

★ Ulcerative colitis is presence of ulcers in colon with rectal bleeding.

★ Crohn's disese involves inflammation of mouth up to anus.

★ Irritable bowel syndrome is blockage of large intestine.

★ Diverticular disease is herniation of colon.

Liver

★ The liver converts ammonia to urea.

★ The liver breaks down haemoglobin, creating metabolites that are added to bile as pigment (bilirubin and biliverdin).

★ The liver also produces insulin-like growth factor 1 (IGF-1), a polypeptide protein hormone that plays an important role in childhood growth and continues to have anabolic effects in adults.

★ Liver can synthesize can 2 g/day of cholesterol.

★ Vitamin E prevents haemolysis of RBC by oxidizing agents and liver injury by carbon tetrachloride.

★ Deficiency of selenium leads to liver necrosis.

★ Diseases of liver are cirrhosis, hepatic coma, viral hepatitis, necrosis of liver.

★ In hepatic coma, branched chain amino acids valine, leucine and isoleucine are reduced while, aromatic amino acids like tryptophan, phenylalanine and tyrosine are increased.

★ HCV is one of several viruses that can cause hepatitis.

★ Ascites is accumulation of abnormal amounts of fluid in abdomen.

Kidneys

★ Human kidney is 10 cm in length, 5 cm wide and 3 cm thick.

★ There are two kidneys, one on each side of the spine. The asymmetry within the abdominal cavity caused by the liver typically results in the right kidney being slightly lower than the left and left kidney being located slightly more medial than the right.

- ★ The upper parts of the kidneys are partially protected by the eleventh and twelfth ribs and each whole kidney and adrenal gland are surrounded by two layers of fat. Each adult kidney weighs between 125 and 170 grams in males and between 115 and 155 grams in females. The left kidney is typically slightly larger than the right.
- ★ The kidney is divided into two main areas: a light outer area called the renal cortex and a darker inner area called the renal medulla.
- ★ Renin is released from kidneys in response to low blood pressure, converts angiotensin I to angiotensin II, also stimulates aldosterone release from adrenal glands leading to salt and water conservation.
- ★ There are three steps in urine formation: glomerular filtration, tubular reabsorption and tubular secretion.
- ★ The normal urine pH is 6 (4.5-8.2) and its yellow colour is due to urochrome.
- ★ Diseases of kidney are: Glomerulonephritis, Nephrotic Syndrome, Acute Renal Failure, Chronic Renal Failure, Urolithiasis.
- ★ Acidosis is the major problem in Chronic Renal Failure with development of azotemia and clinical uraemia syndrome.
- ★ In Chronic Renal Failure, the activation of Vitamin D and the action of parathyroid hormone in controlling the serum calcium and phosphorus level can not proceed at normal level.
- ★ Diuretics increases the rate of urination.
- ★ Dialysis is a procedure that is a substitute for many of the normal duties of the kidneys. Dialysis can allow individuals to live productive and useful lives, even though their kidneys no longer work adequately. It can be peritoneal and hemodialysis.
- ★ In peritoneal dialysis, an osmotic pressure gradient is applied by the addition to the dialysis fluid of an osmotic agent which will "suck" fluid from the blood.
- ★ In hemodialysis, a machine filters harmful waste and excess salt and fluid from blood.

- Kidney stones can be of calcium, oxalate, struvite xanthine and cystine.
- We can see calcium and struvite stones by x-rays.
- Oxalate stones can also arise due to excess fat in colon precipitating calcium and increasing oxalate solubility.
- Xanthine is purine base produced by purine degradation.
- Only struvite stones arises from alkaline urine.
- Struvite stones are caused by microbial infection by *Klebsiella, Providencia, Serratia* and *Proteus.*
- Lithotripsy or litholopaxy is treatment of crushing kidney stone through vibrations.
- Urea splitting organisms like *Proteus, Klebsiella, Pseudomonas, Enterobacter.*
- Nanobacteria causes calcium phosphate stones in astronauts as discovered by NASA.
- Gout is caused by excess uric acid in blood.

Heart

- The heart is located between the lungs behind the sternum and above the diaphragm. It is surrounded by the pericardium.
- The walls of the heart are composed of cardiac muscle, called *myocardium.*
- Ischaemic heart disease occurs due to deficient blood supply.
- Excess cholesterol in blood gradually leads to deposition under lining of blood vessels. This is known as atherosclerosis in which blood vessels are narrowed and hardened. The coronary arteries supplying blood to the heart are affected and coronary heart disease results.
- HDL (High Density Lipoproteins) are good because they play a role in reverse transport of cholesterol from tissues throughout the body back to liver for its conversion to bile acid or excretion as biliary cholesterol. Also HDL has role as an antioxidant in vessel wall.

- ★ Linoleic acid prevents the accumulation of cholesterol in blood serum and walls of blood vessels, plays a key role in cholesterol transport.
- ★ Blood pressure can be categorized as:

Mild (Diastolic Pressure)	90-104mm Hg
Moderate	105-119mm Hg
Severe	120-130mm mm Hg
Normal	80 mm Hg Diastolic Pressure and 120 Systolic Pressure mm Hg

- ★ The pressure in general circulation is about 120/80 mm Hg, in the pulmonary arteries, it is only around 25/15 mm Hg.
- ★ Blood pressure is affected by calcium, magnesium, potassium and sodium. More thick the blood, more chances of high BP.
- ★ A stroke occurs when a blood clot blocks a blood vessel or artery, or when a blood vessel breaks, interrupting blood flow to an area of the brain. When a stroke occurs, it kills brain cells in the area surrounding where the clot or breakage occurs.

Diabetes

- ★ The islets of Langerhens of pancreas contain 3 types of cells: Alpha, Beta and Delta.
- ★ Insulin is a peptide hormone produced in the body by the pancreas.
- ★ Insulin is produced from endocrine part of pancreas which is an exocrine gland by B cells.
- ★ Insulin is released in two phases: Rapid phase (for 10 minutes of after food intake) & slow phase (independent of sugars).
- ★ Release of insulin is inhibited by noradrenaline that causes increase in blood sugar and thus prevents insulin shock.
- ★ C-peptide & Amylin are two other hormones produced by beta cells of Langerhans.
- ★ Insulin converts glucose into glycogen or fats & prevents hyperglycaemia.

Difference between Diabetes Types I and Type II

Identifying Feature	*Type 1*	*Type 2*
Age at diagnosis	Usually <40	Usually >40
Weight	Usually thin with recent weight loss	Usually obese
Condition at diagnosis	Usually moderately to severely ill	May show only mild or no symptoms
Acute complications	Ketoacidosis	Nonketotic, hyperosmolar, hyperglycemic coma
Plasma insulin/C-peptide	Low to none	Inappropriately low for plasma glucose
Response to therapy	Responsive	Variable
Insulin	Unresponsive	Responsive
Sulfonylurea		

- ★ Gestational diabetes is fully treatable but requires careful medical supervision throughout the pregnancy. About 20 per cent –50 per cent of affected women develop type 2 diabetes later in life.
- ★ Normal blood glucose level is <80-120mg/100ml.
- ★ Random blood sugar level is >200mg/dl.
- ★ In diabetes, the presence of glucose in urine is up to 6 per cent.
- ★ HLA class II genes are responsible for diabetes.
- ★ Insulin shock is low blood sugar level up to 45mg/dl.

Diagnostic tests for diabetes:

Test Used	*Blood Samples taken and Analyzed from Plasma Glucose*
Fasting plasma Glucose (FPG)	After fasting (10-14 h without food)
Oral glucose tolerance test (OGTT)	At fasting and then 2h after a 75 g oral glucose drink (in children 1.75g/kg ideal body weightg/kg ideal body weight up to 75 g; in pregnant women a 100-g glucose dose is used and fasting, 1-h, and 3-h samples are taken as well as 2-h samples)
Postprandial plasma glucose (PPG)	2 h after a meal: this measurement is often mistaken for the OGTT; however, it is not clinically reproducible and therefore is less accurate than the OGTT
Random plasmaglucose (RPG)	Without regard to time of last meal

* Abnormal blood glucose during pregnancy turns the pancreas of fateus to secrete more insulin which can cause hypoglycemia in baby of after birth.
* Glycosylated haemoglobin level in normal person is 4-7 per cent ad in diabetes, it is 8-18 per cent.
* Pre-diabetes is a term that identifies those who have some degree of impaired glucose homeostasis, but who do not have diabetes.
* Acetoacetic acid, acetone and hydroxyl butyric acid are known as ketone bodies.
* Hormone insulin lowers the blood level while glucagan, steroids, epinephrine, ACTH, thyroxin, growth hormones increase blood sugar level.
* Epinephrine is given to diabetic patient to overcome insulin shock.
* Somatostatin produced by the islets of Langerhans. It can suppress both glucagon and insulin when needed.

Overweight & Obesity

* Parameters to diagnose obesity are body weight, total body fat, skin fold measurement.

Classification of BMI (kg/m^2) (WHO, 2004):

Underweight	< 18.5
Severe thinness	<16.00
Moderate	16-16.99
Mild	17-18.49
Normal	18.5-24.99
Overweight	≥25.00
Pre-obese	25-29.99
Obese	≥30
Grade I obesity	30-34.99
Grade II obesity	35-39.99
Grade III obesity	≥40

* Obesity gene is called Ob.
* Bariatric surgery is the last option for obese patients who are not able to control their weight.

* Mutation in the human gene for the B3 receptors in adipose tissue, involved in lipolysis and thermogenesis markedly increase the risk of obesity

Other Metabolic Fats

* Progesterone is a steroid hormone.
* DOTS is Directly Observed Treatment Short Course.
* DPT stands for Diptheria, Tetnus and Pertusis.
* DHA is Decosahexanoic acid which is important for neural division in brain.
* TRH stimulates cells in the anterior pituitary gland to release thyrotropin (also known as TSH) into the circulation, where it binds with receptors on cells in the thyroid gland, thereby stimulating the release of thyroid hormone.
* Cigarette smoking inhibits bicarbonate ion production and increases gastric emptying.
* Human body contain about 15L of Extracellular fluid (ECF) composed of blood plasma and interstitial fluid, having pH 7.4.
* ECF contains electrolytes such as sodium, potassium, calcium, chloride ions and bicarbonates.
* In hyperthyroidism there is excess production of T_4 (Thyroxine) and T_3 (Triidothyronine) which can lead to osteoporosis due to excess resorption of calcium and phosphorus from blood.
* Leukemia is blood cancer or cancer of bone marrow. There is lack of blood platelets.
* Brachy therapy is used for cancer.
* Parathyroid hormone is secreted by endocrine glands which are four in number and they maintain calcium and phosphorus level.
* Blood is composed of plasma, WBC, RBC and platelets.
* Human body excretes about 1.5 L of urine in a day.
* Cobalt is required for optimum utilization of low doses of iodine.
* In osteomalacia, anatomic bone volume remain unchanged but there is a feeling of pain in bones and calcium is low.

- ★ In osteoporosis, bone size remain unchanged and is asymptomatic until fracture occurs. Back pain is there and causes loss of height. Body has low calcium and phosphorus.
- ★ Glucuronic acid (UDP) is required for conjugation of bile salts, steroid hormone metabolites and xenobiotics to render them water soluble for excretion.
- ★ Bifedobacterium is bacteria present in child after it is fed on breast milk and it provides better digestibility to child.
- ★ Human body has 4-5 lit of blood.
- ★ TSH is Thyroid Stimulating Hormone and TSH is Thyroid releasing hormone.
- ★ When a burn injury involves deeper layers of dermis it is second degree burn.
- ★ Squint in eyes clinically is known as strabismus.
- ★ Involuntary movement of eyes from one side to another is known as nystagmus.
- ★ Sate of continuous contraction of muscle is known as spasticity.
- ★ Convulsions means sudden shaking of body.
- ★ The patient with arthritis should keep weight within the normal range. Excess body weight increases the stress on joints, especially weight-bearing joints such as the knee and hips.

11

Dietary Modifications

In addition to drug therapy and surgical in spite of medicines and surgical procedures, controlled intake of nutrients plays a very important role in complete and fast recovery of a patient. Well planned diet that incorporate the principles of adequacy, balance and moderation can be nutritious and healthful. Lets go through a fast review on diet planning in normal and diseased conditions.

* A well balanced diet should have 60-65 per cent carbohydrates, 20-30 per cent fat and 10-15 per cent protein.
* In a balanced, ratio of energy distribution from carbohydrate, protein and fat would be 7: 1: 2.
* Not more than 20 percent calories should come from fats.
* Most effective combination to achieve supplementary effect is 4 parts of cereals protein plus 1 part of pulse. In terms of grains it will be 8 parts of cereals and 1 part of pulse.
* Flour should not be sieved for Chapathi as it reduces its bran content.
* 2 to 3 servings of pulses should be taken every day.
* The daily diet of an adult should contain at least 40 g of dietary fibre.
* Tea and coffee should be avoided with meals as they interfere with iron absorption.

- ★ Cooking in cast iron pans and skillets can help to increase the iron content of food.
- ★ Increasing intake of Vitamin-C rich foods, such as orange juice may enhance absorption of non-heme iron.
- ★ One-third of nutritional requirement in terms of calories and protein should be met by lunch or dinner.
- ★ TPN is Total Parenteral Nutrition means providing nutrients through veins.
- ★ Enteral nutrition can be through nasogastric, nasoduodenal and nasojejunal routes.
- ★ Clear fluid diet is that which has no residue. *e.g.* Dal water, barley water.
- ★ Full fluid diet has even no semi solids. *e.g.* Porridges, milk beverages.
- ★ Soft diet has no harsh fibre, low in fat with no seasoning. *e.g.* Dal, finely ground grains.
- ★ In liver diseases, medium chain triglycerides (MCT) can be given as they can be digested and absorbed in absence of bile salts *e.g.* coconut oil.
- ★ A high calorie, high protein, high carbohydrate, moderate or restricted fat, high vitamin diet helps in regeneration of liver and helps to prevent the formation of ascites.
- ★ In hepatic precoma and coma, protein is withheld as liver can not metabolise it. Only high carbohydrate diet is recommended.
- ★ In hepatic coma, low protein diet is given.
- ★ For patient with gastric ulcers, cream or milk is suggested but their quantity should be limited as the fat present in them can lead to gain in weight.
- ★ HDL is known as good cholesterol.
- ★ The proportion of PUFA to SF should be 8:1.
- ★ Arachidonic acid is precursor of prostacyclin which is a vasodialator.
- ★ Omega 3 *i.e.* linolenic acid has better cholesterol control.

- ★ WHO recommends 6g of salt from mixed food sources/day for a normal person.
- ★ DASH is dietary approach to stop hypertension.
- ★ Isoflavones present in soya bean are genistein and diadzein and has anti thrombatic activity.
- ★ Fenugreek seeds are rich in fat, protein and fibre and saponins in them lower blood lipid levels.
- ★ 1/2-1 clove of garlic is sufficient to reduce cholestrol.
- ★ Diabetics should consume foods with low glycemic index.
- ★ Glycemic index of soya is 25.
- ★ Trigonelline is a fenugreek alkaloid.
- ★ Cinnamon makes body responsive to the hormone insulin, which regulates glucose metabolism.
- ★ 2 serving of soya per day can prevent us from cancer.
- ★ Consumption of stanols and sterol esters slightly reduces the absorption of beta carotene, lycopene and alpha tocopherol.
- ★ Cumin has anti cancer property.
- ★ Following nutrients are known to prevent cancer.

  Beta carotene- Stomach, lung, throat cancer

  Vitamin C- Upper GIT, cervix cancer

  Vitamin E- Oral and pharyngeal cancer

  Selenium- Oesophageal and stomach cancer
- ★ Soy protein contains isoflavones, genistein and daidzein.
- ★ Genistein decreases thrombosis, while Daidzein is a potent antioxidant.
- ★ For weight gain excess calories should be given. An additional 500 cal per day is recommended.
- ★ Vegetarian diets are devoid of cholesterol, low in fat, low in saturated fatty acids and rich in monounsaturated and polyunsaturated fatty acids. People generally lose weight on vegetarian diet.

- ★ Females require less calories than males as their BMR is low and size of body is smaller.
- ★ Foods rich in omega-3 are believed to have an anti-inflammatory effect, which may reduce the pain associated with inflamed joints in arithritis. Omega-3 is found in oily fish, such as sardines, mackerel and salmon.
- ★ Patients with fluid restriction are advised to keep ice cube on lips or to have candy during to combat thrust.
- ★ For the patients with fluid restriction, on thirst it is advised to put ice cubes on lips or to have a candy.
- ★ Acid ash diet is given in struvite, alkaline ash diet is given in uric acid and cystine stones.
- ★ 25g of soya proteins/day helps in maintaining better lipid levels in body.
- ★ Usually for obese, 20 Kcal X Body weight is recommended, while this factor for a normal person is 30 and for undernourished is 40.
- ★ Egg & green leafy vegetables are rich in sodium.
- ★ A handful of nuts can be taken daily. They are good source of MUFA, PUFA & fibre.
- ★ 80 g or higher fat in a day may lead to high blood cholesterol.
- ★ Coconut water contains dialyzable biomolecules which inhibit initial mineral phase & also increases demineralization of performed mineral phase thus prevent kidney stone.
- ★ Banana has oxalic acid breaking properties due to presence of vitamin B_6.
- ★ Barley & horse gram are inhibitors of kidney stones; work as diuretics.
- ★ Lemon has citrate that binds calcium in soluble salt from & inhibits crystallization thus slow down stone formation.
- ★ Pineapple juice certain enzymes which can break fibrins thus prevent stone formation.

12
Food Science and Technology

Introduction

Food Science is branch of science that study the technical aspects of food. It is a study of different chemicals present in food, their chemical and physical properties. Whereas, Food technology, is a branch of science that deals with the actual production process to make foods. Both are the broad areas and brings together multiple scientific disciplines. Both incorporate concepts from fields such as microbiology, chemical engineering and biochemistry.

Key notes on Food Science

For the students of Nutrition and Dietetics, the knowledge of Food Science is of utmost importance. Nutrition and Dietetics gives a broad knowledge of nutrients and their interactions but the study of Food Science acquaints them about the correct proportion and location of nutrients and transformation of a food into different consumer acceptable food products. Since Food Science is a broad area in itself.

The core research areas in Food Science and Technology involves:

- Toxin removal, preservation, easing marketing and distribution tasks and increasing food consistency through long distance transportation.

★ Reduction of incidences of food borne disease.

★ Positive and negative interaction between micro organisms and disease and food preservation.

This chapter is highlighting imporant ones

★ Product development, choice of packaging material, shelf life studies as well as chemical and microbiological testing.

★ Waxy cereals contain high proportion of amylopectin.

★ The gelatinization dramatically increases the digestability of starch.

★ The recrystallization of starch is referred to as retro gradation and may reduce the digestibility of the starch.

★ X-rays are used to know the amorphous or crystalline form of starch in food.

★ High amylose barley contain 40 per cent amylase while high amylase maize has 50-80 per cent of amylase.

★ ARF stands for 'Amylase Rich Foods'.

★ Gluten is chief protein of wheat cWomposed of gliadin and glutelins.

★ Roasting is done at a temperature of 140°- 200°C.

★ Hard wheat is employed for making bread because it gives large loaf volume, good crumb while soft wheat is good for biscuit and cake.

★ Amylose is not easy digestible than amylopectin. In this, glucose units are linked in linear way with α -1-4 glycosidic linkage. Branching takes place with α1, 6 linkage occurring every 24 to 30 glucose units resulting in a soluble molecule that can be quickly degraded as it has many end points for enzymes to act.

★ Aghonibora, a rice variety grown in Assam needs not to be cooked. It is parboiled rice which when soaked in water expands and get ready to eat.

★ Glutelin or oryzenin is a protein of rice while the protein of maize is zein.

★ Barley contain hordein as its main protein.

★ Rice proteins:

Water soluble	Albumin	5 per cent
Salt soluble	Globulin	10 per cent
Alcohol soluble	Prolamin	10 per cent
Alkali soluble	Glutelin	80 per cent

- ★ Rice from freshly harvested paddy cooks to a glutinous mass and low swelling property and causes digestive disorders.
- ★ Bajra or barley is known as pearl millet while ragi is known as finger millet.
- ★ Conditioning of wheat is done by drying it to a moisture content of 15-17 per cent for milling.
- ★ Durum wheat is used for pasta formation.
- ★ Saponins present in pulse, produce lather when shaken with water.
- ★ Freshly harvested rice swells two times due to α-amylase content which can be inactivated by its on storage.
- ★ Hominy are the fine fractions of maize obtained after dry milling.
- ★ Malik process and HR conversion process are two types of parboiling methods.
- ★ Saponins and hemaglutenins are present in legumes as antinutritional factors and are toxic to RBC's.
- ★ Maillard reaction occurs at low temperature while carmelization occur at high temperature.
- ★ Onion, garlic, spring onion, leek, fennel are bulb of plants while carrot, turnip, beetroot, radish are roots and potato, yam are tubers of plants.
- ★ Chlorophyll in vegetables is present in chloroplast. Its type A is blue which is present in florets while Type B is yellow green and present in stalk.
- ★ The betalains are plants pigments that are yellow, orange, pink, red and purple in colour. *e.g.* beet roots.
- ★ Anthocyanin give purplish to blue colour to vegetables while Anthoxanthin are responsible for pale yellow. *e.g.* cauliflower, onions, spinach.

- ★ Chlorophyll is fat soluble and when it is subjected to heat it forms pheophytin.
- ★ Chlorophillides formed from chlorophyll gives green colour.
- ★ Anthocyanins are present in black grapes, berries and plums, anthoxanthins in apple, potato and onions while catechin and leuco anthocyanin is present in tea, apple and grapes.
- ★ Broccoli is anticarcinogenic because of presence of sulhoraphane.
- ★ When onion cells are damaged, allinase released form thiosulphates and thiols with lacrymator compounds.
- ★ When garlic is crushed, amino acid allin is enzymatically converted to allicin which converts rapidly to sulphide compounds.
- ★ Pungent flavor in chillies is due to capsaicin.
- ★ Non-climacteric fruit are those which do not ripen after harvesting *e.g.* Citrus fruits.
- ★ Ethylene is responsible for ripening of fruits.
- ★ Calcium carbide is also used for fruit ripening as it form has acetylene but is toxic. Also makes fruit tasteless.
- ★ Artificial ripening of fruits is done by ethephon which decomposes to ethylene.
- ★ The edible "flesh" of the coconut; when dried it is called copra.
- ★ Pectin methyl esterase hydrolyses pectin resulting in loss of gel forming capacity.
- ★ Casein is known as milk protein and in milk is about 82 per cent.
- ★ Gerber's test is used for determination of milk fat.
- ★ Yellow colour of milk is due to the presence of carotenes
- ★ Standardized milk has 4.5 per cent fat and 8.5 per cent SNF while, tonned milk has 7 per cent fat and double tonned milk has 9 per cent SNF.
- ★ SNF is solid not fat means other solids without fat.
- ★ The phosphates test is applied to dairy products to determine whether pasteurization was done properly and also to detect the possible addition of raw milk to pasteurized milk.

- ★ During heating of milk riboflavin is lost.
- ★ Diacetyl is responsible for flavour of butter.
- ★ Whey proteins include lactoalbumin, lactoglobulin, lactoferrin, serum albumin, serum transferring.
- ★ Kumis is mare's milk.
- ★ Difference in temperature after laying of egg by hen causes formation of air at the end of egg.
- ★ Egg white is the common name for the clear liquid (also called the albumen or the glair/glaire) contained within an egg.
- ★ Egg has 12-14 per cent proteins with BV of 96.
- ★ Leghorn breed produces unfertilized egg while country breed produces fertilized egg.
- ★ Egg yolk contain protein livetin.
- ★ Albumen causes foam in egg.
- ★ Flavour in meat is due to natural presence of inosinic acid, glycopetides and amino acids such as glutamic acid.
- ★ Fish has about 20 per cent protein, biological value is 80 and is rich in lysine and methionine. Oysters have good amount of copper, iron and zinc with good bioavailability than plants.
- ★ Peanut butter has a high level of monounsaturated fats and resveratrol.
- ★ Butter has large proportion of SCFA and melts at temperature than beef fat.
- ★ When oil is pressed from oilseeds without heating, such oil is known as cold-pressed or virgin oil.
- ★ Unconventional oils are derived from mango kernal, rice bren, neem & mahua.
- ★ When fats are heated to smoking point they decompose glycerol to acrolein which causes irritation to eyes and nose.
- ★ Lard is fat rendered from adipose tissues of trimmed pork.
- ★ Suet is obtained from shredded adipose tissue from cattle or other ruminents and usually contains little protein.

- ★ Hexane is used in solvent extraction of oil.
- ★ Refining of oil removes suspended particles by filteration, free fatty acids by neutralization and pigments by bleaching.
- ★ By hydrogenation, oleic acid is converted to trans isomer eladic acid.
- ★ HFCS is high Fructose Cane Syrup.
- ★ Glycyrrhizin and thaumatin are non- calorie sweeteners (natural).
- ★ Stevia has sweetness due to stevioside.
- ★ Roasting and baking occurs at temperature of 120°Cand 260°C.
- ★ Pressure cooker works at the temperature of 121°C and pressure of 1.07kg/cm^2 or 15lb/m^2.
- ★ Microwave works on the principle of conduction and convection by heating the water present with in foods.
- ★ Microwave cooks 10 times faster than other methods of cooking.
- ★ The common form of edible mushroom is known as button mushroom which is a fungus named as Agaricus bisporus.
- ★ Coffeol provides flavour coffee. Its bitter taste is because of alkaloids.
- ★ The convenience food could be basically classified into two categories: Shelf – stable convenience food and Frozen convenience food
- ★ Solanin, ricin and opium are anti nutrients present in potatoes, castor and poppy seeds, respectively.
- ★ Pulses, wheat product, veg oil, ground and whole spices, milk, tobaco, walnuts, spices, basmati rice, onions, potato, oils are mandatory to be controlled by AGMARK.
- ★ ISO (International Organization for Standardization) and CAC (Codex Alimentarious Commission) are two international standards.

Key notes on Food Technology

As we have discussed earlier, similar to Food Science, the knowledge of Food Technology is also important for Nutrition and Dietetics students. The idea of day today growth of food industry and of processing steps or techniques is must because no diet or food can be prescribed until the processing technology or its properties, ingredients or chemical additives used are unknown. With this discussion lets start based on some facts on Food Technologies.

- ★ In wheat, ideal moisture should be 12 per cent after drying.
- ★ Moisture content for safe storage should be 13-14 percent and relative humidity should be 70-75 percent.
- ★ Insects can live and reproduce at temperature between 15-35°C.
- ★ About 6-11 percent food grains are lost during post harvest operations.
- ★ N-capaldehyde and n-valeradehyde produced by auto oxidation of lipids are responsible for stale flavour in old rice.
- ★ Grain sprouts when its moisture content exceeds 30-35 percent.
- ★ Methylbromide is a gaseous fumigant used against insects while ethylene bromide is a liquid fumigant.
- ★ Hue indicates colour such as blue or red while chroma is the reflection at a given wavelength.
- ★ Lightness of color is the relation between reflected and absorbed light.
- ★ Beer's law is optical density is directly proportional to concentration of liquid.
- ★ Lambert's law is rate of decrease in intensity with thickness is proportional to intensity of light.
- ★ UV rays have 3 ranges: Expanding (320-400nm), UVA (320-280nm) and UVC (below 280nm).
- ★ Spectrophotometer uses tungsten lamp for measurements in the visible region (360-900 nm) and a hydrogen or deuterium lamp for UV region (200-380nm).
- ★ Spectrofluorimeter uses xenon lamp and colorimeter uses only tungsten lamp.
- ★ Fluorimetery is used for determination of sodium, potassium and calcium only.
- ★ Chromatography seperates molecules on the baiss of size shape, mass, charge, solubility and adsorption. Discovered by Pswett in 1906.

★ Base on mobile phase, chromatography is liquid and gas chromatography. On the basis of stationary phase, it is thin layer, coloumn, gas-liquid and high pressure liquid chromatography (HPLC).

★ On the basis of support, chromatography is TLC (Thin Layer Chromatography), paper chromatography and coloumn while on the basis of interaction, it can be adsorption, partition and ion exchange.

★ Gel filteration chromatography is a powerful tool to separate proteins and also used for separation of viruses, ribosomes, cell nuclei, bacteria.

★ Coloumn materials used in gel chromatography are sphadex, agarose and bio-gel.

★ In gas liquid chromatography (GLC), mobile phase is gas (nitrogen, helium, argon).

★ GLC is used for determination of fatty acids, bile acids, steroids, hormones, sugar and alkaloids.

★ Ion exchange chromatography is based on electrostatic attraction of oppositely charged ions on polyelectrolyte surface.

★ HPLC (High Pressure Liquid Chromatography) is used for determination of fungicide, non-volatile organic metabolites, ascorbic acid, niacin, caffeine and phenyalanine.

★ ELISA is Enzyme Linked Immunosorbant Assay. It can be sandwitch, competitive and indirect.

★ Atomic absoption spectrophotometry determines the concentration of substance by measurement of adsorption of radiation by atomic vapour of an element.

★ Atomic absorption apectrophtometery is used for determination of trace elements.

★ Flame photometery determines the concentration of an element in a solution by measurement of emission of light by atoms of elements in flame.

★ When a particle revolves around an axis, a force is developed which act away from axis of rotation called centrifugal force.

- ★ Principle of refractometer is when a ray of electromagnetic radiation strikes ion flat surface at an angle, the ray may bent upward or downward refracted.

Food Quality and Preservation

- ★ TQM is Total Quality Management. It is an integrated organizational approach in delighting customers by meeting their expectations.
- ★ HACCP is Hazard Analysis Critical Control Point. It is a general approach to problem solving as well as ensuring product safety.
- ★ The quality of butter is measured by consistometer, of cheese by rotating and oscillating viscosimeter, of jelly by jelmeter and of candy by oswalt viscosimeter.
- ★ The quality of fruits and vegetables can be assessed by penetrometer, compressimeter, sheer testing device and by energy measuring meter.
- ★ The dough quality can be assessed by recording dough mixer, brabender farinograph, mixograph, brabender extensiograph, research extensiometer, alveograph and amylograph.
- ★ In baking industry, amylase enzyme helps to accelerate fermentation It increases loaf volume, improvement of crumbs and softness.
- ★ Lipooxygenase activity causes direct oxidation of lipids containing cis, cis 1-4 petadiene to hydroperoxides in presence of cupric and ferric ions. It causes rancidity, acidity, discolouration, soapiness and off flavor.
- ★ Rancidity can be absorptive, hydrolytic and oxidative.
- ★ Rancidiy of fats occur due to auto oxidative deterioration, lipolysis and lipoxidation.
- ★ Enzymatic browning occur due to polyphenol oxidase or orthodiphenol oxygen oxido reductase EC 1.10.3.1 and causes formation of melanine from tyrosine.
- ★ In sensory evaluation of foods, trained panel members should be 5-10, semi trained should be 25-30 and untrained panel members should be 100 in number.
- ★ Sensory evaluation tests:

1. Difference test

   Paired comparison test: Sample size-2

   Duo trio test: Sample size-3 (two identical, one different)
   Triangle test: Sample size-3 (two identical, one different)

2. Rating test

   Single sample test

   Two sample test

   Multiple sample test

3. Sensitivity threshold test

4. Descriptive test

* Radiation processing can be used for insect disinfestations, inhibition of sprouting, delay in ripening, destruction of microbes and for elimination of pathogen and parasites form from foods and known as Irradiation or cold sterilization.

* In irradiation, cobalt-60 and cesium-137 is used as source of gamma rays while x rays of 5mev and electron beam of 10mev are used.

* Radiation dose is measured in terms of Grey/KGrey and its upper use limit is 24KGrey of KGy.

* The upper limit of use of electron beam is 10mev.

* In Lasal Gaon, Mumbai, onion treatment radiation plant is there while for spices it is located at Vashi Navi, Mumbai.

* Microwave oven was introduced in 1952 in America by Raytheon company and was invented by Germany during 2nd world war to support their Navy.

* Microwaves in oven act on principle of interference, reflection, refraction and polarization. It uses microwaves of wavelength of 0.5-50cm which has frequency of 6×10^8 to 3×10^{10} Hz. Microwave oven contains device magnetron.

* Magnetron in microwave oven, converts electric energy into electro magnetive force (EMF) which causes dipole rotation and generates heat. Amount of energy extracted in microwave oven depends upon electric field strength, electric property of food, food and microwave's frequency.

- ★ Heat produced in the food being microwaved is distributed to all parts of food by conduction- conduction phenomenon.
- ★ Only glass and ceramics can be used for microwave oven.
- ★ High hydrostatic pressure, ultrasound, pulsed electric field, light pulses and oscillating magnetic field are some non thermal methods that preserve foods.
- ★ The principles of preservation are:
    - Prevention or delay of microbial decomposition.
    - Prevention or delay of self decomposition.
    - Prevention or delay of purely chemical reactions.
    - Prevention of damage by insects-pests.
- ★ Reduction of water activity is the principle of drying.
- ★ Spoilage is characterized by rancidity and acidity due to production of free fatty acids such as butyric acid, capric acid and methyl ketones.
- ★ The enzyme that causes browning of fruits and vegetable is polyphenol oxidase. It works at pH of 5-7 and temperature of 25-35°C.
- ★ In 1987, Labeza introduced active packaging whose types are: oxygen scavenging, antimicrobial, ethylene control, moisture control, gas permeability and odor remover.
- ★ Wort is the clear liquid obtained after fermentation with hops during beer formation. It contains all fermentable sugars.
- ★ Sherry is an appetizing wine while champagne is a sparkling wine.
- ★ Sparkling wines do not contain carbon dioxide while sparkling wines are made effervescent by secondary fermentation.
- ★ Blanching is a treatment of fruits and vegetables with boiling water or steam for short period followed by cooling.
- ★ During canning syrup or brine can be added at 79-82°C with space left inside 0.3-0.47cm and then exhausting of cans is done to remove air and better retention of vitamin C, for avoiding corrosion of tin plate and to prevent discoloration of product. It is done at 82-87°C for 5-25 minutes.

Food Microbiology

- ★ D-value is Decimal Reduction Time and can be defined as time required to kill 90 per cent of micro organisms or spores in a sample at a specific temperature. Another definition is time required for the line to drop one log cycle.
- ★ Z -value is increase in temperature to reduce D value by 1/10 of its value.
- ★ F -value is time in minutes at a specific temperature to kill a population of cells or spores.
- ★ 5D or 12 D is the enough heat required to reduce 10^5 or 10^{12} spores to one spore/ml
- ★ Curd is made by *Streptococcus lactis* and *Streptococcus acidophilus* while yoghurt is made by *Streptococcus thermophilus* and *Lactobacillus acidophilus*.
- ★ Wine is made from grapes variety *Vitis vinifera.*
- ★ Lager beer is made by yeast that settles to the bottom of fermentation tank.
- ★ Nira is fermented beverage from palm tree while rum is produced from molasses.
- ★ Cider in India and Europe is fermented while in US it is clarified and unfermented.
- ★ Hops are unfertilized dried flowers of female wine (*Humulus lupulus*) added in beer to give astringent and bitter flavor. They also inhibit undesirable micro organism growth.
- ★ Ergotism or St Anthony's fire is caused by Calviseps pupurea.
- ★ Aflatoxins are produced by Aspergillus flavus.
- ★ Food infection is caused by *Salmonella, E coli, Llisteria, Yerseni, a Steptococcus, Shigella* and *campylobacter.*
- ★ J. Deny in 1894 first studied staphylococcal food poisoning which is caused by *Staphylococcus, Clostridium* and *Bacillus.*
- ★ *Clostridium botulinm* and *Staphylococcus aureus* are classical examples of food poisoning.
- ★ *Salmonella* causes typhoid.

- ★ Air can be treated to make it microbe free by rain, sunshine, laminar flow, formaline, hypochlorite and by UV rays.
- ★ In milk, lectenins and lactoferin are present which act inhibitor of enzymes.
- ★ TDP is Thermal Death Point and is defined as lowest temperature at which a microbial suspension is killed in 10 minutes.
- ★ TDT is Thermal Death Time and defined as shortest time to kill all micro organisms in a microbial suspension at specific temperature and under defined conditions.
- ★ Bergey's manual is the main resource of determining identity of bacteria species.
- ★ Bio K plus is a probiotic and can be taken in lactose intolerance as it is free from dairy products.
- ★ The example of hard cheese is swiss cheese, of semi hard is Roquefort and of soft cheese is limburger and camembert.
- ★ During ripening of cheese mico organisms are applied on interior to make it hard and on inner side to make it soft cheese.
- ★ Ripening of cheese takes place at 10°C with high humidity.
- ★ There are some edible micro organisms like *Chlorella* and *Spirulena* (Algae), *Bacillus* and *Acinetobacter* (Bacteria), *Agaricus* and *Apergillus* (Fungi), *Sachromyces* and *Candida* (Yeast).
- ★ Water activity required for growth of yeast is 0.88, for mould 0.80 and for bacteria is 0.91.
- ★ The reason of food spoilage are microbial activity, action of enzymes, chemical reactions and physical change.
- ★ Pasteurization is done to kill two bacteria: *Mycobacterium tuberculosis and Coxiella burnetti.*
- ★ $KMnO_4$ or NaOH is added to water before distillation in order to oxidize steam volatile compounds.
- ★ Autoclaving works on the principle that water boils when its vapour pressure become equal to pressure of surrounding atmosphere.
- ★ The pressure used for autoclaving is 15lb/sq inch or done at 121°C for 15-20 minutes while for compounds containing sugar we use temperature of 115°C for 20 minutes to prevent their caremalization.

Table 12.1: Nutritional Improvement of Food

Method	*Definition*	*Advantage*	*Example*
Fortification	Addition of more than one essential nutrient to a food whether or not that nutrient is normally contained in the food	Improves nutrient intake levels of target population which are at risk of micronutrient deficiency. Prevention or correction of demonstrated deficiency of one or more nutrients in the population or specific population groups.	Iodized salt, Golden rice.
Enrichment	Restoration of vitamins and minerals lost during processing stage	Addition of lost nutrients to the refined product. Eg. thiamine, niacin, riboflavin, iron etc	Wheat grain is a good source of fiber and nutrients but after refinement of wheat it will mainly a source of calories.
Complementation	Two partially complete proteins are given together.	A deficiency of an amino acid in one can be made-up to an adequate level in another, if both are consumed together.	Balahar which is a combination of cereal and pulses provided in aaganvadi and schools in mid day meal. Other examples are khichdi, idli, dosa, dal-rice.
Fermentation	Metabolic process in which a micro-organism converts carbohydrate present in food, such as starch or a sugar into an alcohol or an acid.	Preserves the food, increase the shelf life of product, improve digestion, creates beneficial enzymes, produce vitamins such as B complex and vitamin C, increases flavour, omega -3 fatty acids and some strains of probiotics	Bread, cheese, yoghurt, sauerkraut, kimchi etc

Method	Definition	Advantage	Example
Germination	Process of soaking seeds in water that causes the complex carbohydrates in seeds to break down into simple sugars and the proteins to break down into amino acids	Sprouted rice reduces attack of common allergens and reduces the risk of cardiovascular, sprouted brown rice fights diabetes. Sprouted rye increases and protects folate.Sprouted buckwheat protects from fatty liver and its extract decreases blood pressure, increasesvitamins (usually B and C) and other nutrients	Rice, pulses, lentils etc.
Parboiling	Process in which rice is boiled before milling under specific temperature and pressure	The brokens in the milled rice is minimized. Less infestation Better cooking quality More B-vitamins than milled raw rice. Less loss of B-vitamins during washing and cooking. A bran higher in oil content (about 25-30 per cent oil)	Paddy
Supplementation	Process of adding complete proteins to partially complete/ incomplete proteins.	Supplementation of limiting essential amino acid with protein concentrates high in amino acids can be done.	Kheer, corn flakes with milk, bread-omellete, porridge with milk etc.

Impact of Cooking on Food

★ Cooking involves subjecting food to heat and when food is subjected to heat then some chemical changes occur in food and to the nutrients as well. Certain changes are beneficial and others are not. Beneficial changes include destruction of microbes, improvement in texture etc. Cooking can also lead to loss of nutritive values but as it improves the shelf life of food so cooking and processing of food is beneficial. The chapter outlines the effect of cooking on food:

- Cooking increases the palatability of food. It make the food more palatable and edible. For example raw sweet potato is inedible but cooking (boiling or roasting) makes it edible and palatable.
- Some basic staple foods such as dry legumes and whole grains are not in edible form when harvested and these products must be rehydrated to soften the texture.
- Although meat, fish and poultry are eaten raw in some population but cooking improves the aesthetical appeal and increases the palatability. Also cooking improves the hygiene quality of food and heat destroys the microorganisms. However the extent of microbe destruction depends upon the time and temperature relationship.
- Digestibility and nutritive value in some cases is increased by cooking. Starch and cooked grain products and legumes become more readily available to digestive enzyme than in compact raw starch granules.
- Some anti nutrient factors like trypsin inhibitors, saponins, haematoglutenins etc. are present in pulses and legumes. Soaking, germination and cooking removes the antinutritive factors.
- Apart from removing antinutritive factors cooking also bring about some decrease in nutritive value as well. Heat sensitive water soluble vitamins (B-vitamins and vitamin-C) get destroyed during cooking.
- Cooking and processing of food improves the keeping quality of food as cooking above 70 degee C will kill most of the microbes however pathogenic micro-organism requires a little high

temperature. Also, cooked food should be maintained away from temperature danger zone (7-65 degree C) to inhibit the growth of micro-organism.

- Cooking also makes it possible to show creativity and makes many dishes from same raw material. Thus, it adds variety to the diet. For example, rice can be used in making kheer, pulav, boiled rice, biryani, etc.
- Processing food also affect flavor, colour and texture of food. New flavours and improvement of texture is observed during baking, roasting, frying etc. for example new flavours are developed during making of bread, caramel and cooking meat. However, at the same time some flavours may be lost or undesirable flavours may develop.
- Texture of vegetable softens as fibers of vegetables and connective tissues of meat are tenderized.
- Colour changes also occur during cooking. Bright green vegetable turns dull when food is cooked in presence of acid/ alkali whereas short blanching actually enhances the colour of fresh green peas. In case of baked products, light brown colour and crust formation occurs which improves the appeal of the food.

13

Food Additives

Introduction

Food additive is any substance the intended use of which directly or indirectly becoming a component of or otherwise affecting the characteristic of any food and which is safe under the condition of its use. Food additives are substances which are added to food which either improve the flavor, texture, colour or chemical preservatives, taste, appearance or function as processing aid. Food additives are non-nutritive substances added intentionally to food generally in small quantities to improve its appearance, flavor, texture or storage properties.

Need for Food Additives

- For safe storage, transportation, distribution or processing.
- To maintain food distribution.
- For continuing the convenience food revolution.
- Many of these chemical additives can be manufactured so that foods can be "fortified" or "enriched".
- Potassium iodide, for instance, added to common salt can eliminate goiter, enriched rice or bread with B-complex vitamins can eliminate pellagra and adding vitamin D to cow milk prevents rickets.

Classification

There are thousands of food additives. Some important are discussed below:

1. Anti-oxidants

- ★ An anti-oxidant is a substance added to fats and fat-containing substances to retard oxidation and thereby prolong their wholesomeness, palatability and sometimes keeping time.
- ★ Some anti-oxidants used in foods are Butylated Hydroxyanisole (BHA), Butylated Hydroxytoluene (BHT), Propyl Gallate (PG) and Teriarybutyl Hydroquinone (TBHQ), which are all phenolic substances. Thiodipropionic acid and dilauryl thiodipropionate are also used as food anti-oxidants.

2. Chelating Agents

- ★ Chelating agents are not anti-oxidants. They serve as scavengers of metals which catalyze oxidation.
- ★ Examples are EDTA, dimercaprol and sodium gluconate.

3. Colouring Agents

- ★ These include colour stabilizers, colour fixatives, colour fixatives, colour retention agents, etc.
- ★ A number of natural food colours extracted from seeds, flowers, insects and foods are also used as food additives. One of the best known and most widespread red pigment is bixin, derived from the seed coat of Bixa orellana, the lipstick pod plant of South American origin.
- ★ Annatto has been used as colouring matter in butter, cheese, margarine and other foods.
- ★ Another yellow colour, a carotene derived from carrot, is used in margarine.
- ★ Saffron has both flavouring and colouring properties and has been used for colouring foods.
- ★ Turmeric is a spice that gives the characteristic colour of curries and some meat products and salad dressings.

- ★ A natural red colour, cochineal (or carnum) obtained by extraction from the female insect (*Coccus cacti*), grape skin extract and caramel, the brown colour obtained from burnt sugar, are some natural colours that are used as food additives.

4. Curing Agents

- ★ These are additives to preserve (cure) meats, give them desirable colour and flavor, discourage growth of micro-organisms and prevent toxin formation.
- ★ Sodium nitrite has been used for centuries as a preservative and colour stabilizer in meat and fish products.
- ★ The nitrite, when added to meat, gets converted to nitric oxide, which combines with myoglobin to form nitric oxide myoglobin (nitrosyl myoglobin), which is a heat-stable pigment.
- ★ The curing also contributes flavor to the meat. In addition, nitrite curing inhibits the growth of *Clostridium* and *Streptococcus* and also lowers the temperature required to kill *C.botulinum*.

5. Emulsifiers:

- ★ Emulsifiers are a group of substances used to obtain a stable mixture of liquids that otherwise would not or would separate quickly.
- ★ They also stabilize gas-in-liquid and gas –in-solid mixtures. They are widely used in dairy and confectionery products to disperse tiny globules of an oil or fatty liquid in water.
- ★ Emulsifying agents are also added to margarine, salad dressings and shortenings.
- ★ Eg. Lecithin, mono glycerides of fatty acids and sorbin esters

6. Flavours and Flavour Enhancers

- ★ The agents responsible for flavor are esters, aldehydes, ketones, alcohols and ethers. Typical of the synthetic flavor additives are amyl acetate for banana, methyl anthranilate for grapes, ethyl butyrate for pineapple, etc.
- ★ One of the best known, most widely used and somewhat controversial flavor enhancers is monosodium glutamate (MSG), the sodium salt of the naturally occurring amino acid glutamic acid.

- ★ Yeast extract has the same flavor enhancing property as MSG. It is ten times more powerful than MSG.

7. Flour Improvers

- ★ These are bleaching and maturing agents. Usually they both bleach and "mature" the flour.
- ★ These are important in the flour milling and bread-baking industries.
- ★ Chemical agents used as flour improvers are oxidizing agents, which may participate in bleaching only, in both bleaching and dough improvement, or in dough improvement only.
- ★ The agent that is used only for flour bleaching is benzoyl peroxide. This does not influence the quality of dough.
- ★ Materials used both for bleaching and improving are chlorine gas, chlorine dioxide, nitrosyl chloride and nitrogen di and tetraoxides.
- ★ Oxidizing agents used only for dough improvement are potassium bromate, potassium iodate, calcium iodate, and calcium peroxide.

8. Humectants:

- ★ Humectants are moisture retention agents. Their functions in foods include control of viscosity and texture, bulking, retention of moisture, reduction of water activity, control of crystallization and improvement or retention of softness.
- ★ Some of them are propylene glycol glycerol, and sorbitol and mannitol .

9. Anti-caking Agents:

- ★ Anti-caking agents help prevent particles from adhering to each other and turning into a solid chunk during damp weather. Thus help in free flowing of salt and other powders. *e.g.* Calcium silicate, magnesium carbonate, kaolin, bantonite and sodium alumino silicate.

10. Leavening Agents

- ★ Leavening agents produce light fluffy baked goods. Originally, yeast was used almost exclusively to leaven baked products. It is still an important leavening agent in bread making.

11. Non-nutritive Sweeteners

- ★ The first synthetic sweetening agent used was saccharin (sodium ortho benzene sulphonamide or the calcium salt), which is about 300 times sweeter than sucrose in concentrations up to the equivalent of a 10 per cent sucrose solution.
- ★ Acesulfame K is used in baked goods, chewing gum, gelatin desserts, and soft drinks. It is about 200 times sweeter than sugar.
- ★ Aspartame is used in "Diet" foods, including soft drinks, drink mixes, gelatin desserts, and low calorie frozen desserts. Aspartame is producedfrom two amino acids: aspartic acid and phenylalanine and is 180 times sweeter than sucrose.

12. pH Control Agents

- ★ These include acids, alkalis and buffers. They not only control the pH of foods but also affect a number of food properties such as flavor, texture, and cooking qualities.

13. Preservatives

- ★ A preservative is defined as any substance which is capable of inhibiting, retarding or arresting, the growth of micro-organisms, of any deterioration of food due to micro-organisms, or of masking the evidence of any such deterioration.
- ★ The compounds used as preservatives include natural preservatives, such as sugar, salt, acids, etc, as well as synthetic preservatives. Benzoic acid, Na-benzoate, sorbic acid , propionic acid and its salts, acetic acid

14. Stabilizers and Thickeners

- ★ These compounds function to improve and stabilize the texture of foods, inhibit crystallization (sugar, ice), stabilize emulsions and foams, reduce the stickiness of icings on baked products, and encapsulate flavours.
- ★ Substances used as stabilizers and thickeners are polysaccharides, such as gum Arabic, guar gum, carrageenan, agar-agar, alginic acids, starch and its derivatives, carboxy methylcellulose and pectin. Gelatin is one non- carbohydrate material used extensively for this purpose.

15. Other Additives

- Clarifying agents like bentonite, gelatins, synthetic resins (polyamides and poly vinyl pyrrolidone) are used to clear the haziness in juices.
- Firming agent: Keep the fruits/ vegetable tissues firm/ crisp, e.g. Calcium alum etc., used in fruits and vegetable candies.
- Fungistatic agent: Sorbic acid:, used in bread, cheese, cake etc. used in food packaging paper.
- Antibiotics: Nisin and Chlorotetra cycline, used in dairy products.

Advanced findings on food additives:

- *When food additive is categorized as GRAS it means 'Generally Recognized as Safe'.*
- *The E numbers are given by European union to identify the chemical additive added in foods.*
- *E 100-199 - Colour, E 200-299 - Preservatives, E 300-399 - Acids and Antioxidants, E 400-499 - Thickeners, Stabilizers and Emulsifiers, E 500-599 - Anticaking Agents, E 600-699 - Flavours and Flavor Enhancers, E 900-999 - Miscellaneous and E 1000-1999 _Additional Chemicals.*
- *NOAEL is No Observed Adverse Effect Level. It is divided by 100 to give margin of safety of food additive.*
- *Delane clause in section of food additive states that no additive is deemed safe if it is found to cause cancer.*
- *Aluminum silicate is used in salt as humectant (2 percent by weight).*
- *Curing agent sodium nitrite form nitric oxide which combine with myoglobin and form nitric oxide myoglobin. It cat against Clostridium botulinm. This also forms nitrosamines which are carcinogenic.*
- *The E number of MSG is 621 while of acetic acid is 260.*
- *Colour erythrosine can induce asthma and thyroid diseases while tatrazine can cause asthma.*
- *Preservatives such as nitrates can lead to cancer while benzoic acid can induce asthma.*
- *Antioxidants such as propyl gallate can induce blood disorders while butylated hydroxy toluene (BHA) can be carcinogenic.*

- *MSG is mono sodium glutamate . Also known as Ajino Moto. It acts as flavour enhancer and can cause asthma and hyperactivity in children.*
- *Potassium metabisulphate is a source of sulphur oxide as preservative. It when added to neutral or alkaline media, radicals reacts with acid of juice and form potassium salt the sulfur gas is liberated and from sulphurous acid. This sulphurous acid is an active form of this preservative.*
- *Saccharine an artificial sweetener is known to have carcinogenic properties while aspartame can induce allergic reactions.*
- *Non calorie sweeteners are aspartame, cyclamates, and saccharin while low calorie sweeteners are polyols, xylitol, sorbitol and mannitol.*
- *Suphur gas in foods inhibits growth of E coli and yeast.*
- *Antioxidants like Butylated Hydroxy Toluene works by donating a proton to fatty acid free radical to prevent formation of hydroxy peroxide.*
- *Brominated vegetable oil is that whose density has been increased to that of water by brominaion, causes cloudiness in drinks and protect from light induces reaction but causes cancer.*
- *Preservatives interfere with cell division, permeability of cell memberane and activity of enzymes.*

14

Food Irradiation

Introduction

Scientific studies have given a number of methods of preservation. The recent advancements in the applications of radio-isotopes and radiation technologies in various areas like medicines, industry, agriculture and research have enhanced the peaceful uses of atomic energy and improved the quality of life in many spheres.

Definition

The radiation processing of food or food irradiation can be defined as: "Process of exposing food to ionizing radiation in order to destroy microorganisms, bacteria, viruses, or insects that might be present in the food" It is a controlled application of energy from ionizing radiations to destroy bacteria, pathogens and pests in food and agricultural products greatly reducing the threat of food borne diseases".

Need of Food Irradiation

Following reasons demand use of radiation process to preserve food:

1. Persistently high food losses due to infestation, contamination and spoilage: FAO estimates that up to 25 per cent of world food supply is lost due bacteria and pests.
2. Mounting concern over food borne diseases: According to Centre of Disease Control and Prevention, food borne diseses strikes about 48 million illness, 128,000 hospitalizations and 3000 deaths yearly.

3. Growing international trade in food products must put stiff import standards of quality and quarantine.

Radiation Unit and its limit

- ★ The dose of irradiation is usually measured in a unit called the Gray (Gy).
- ★ Recently in 2003, the Codex Alimentarius Commission (CAC) which is subsidiary to both World Health Organization and Food and Agricultural Organization revised the guidelines for food safety and has raised their limit to 24KGy.

Radiation Treatment

Radiation treatment of food can be divided on the basis of terminology given by an International Group of Microbiologists in 1964.

- ★ Raddappertization or radiation sterilization or commercial sterility used in canning industry and dose level is 30-40 kGy.
- ★ Radicidation equivalent to pasteurization of milk to reduce the number of viable specific non-spore forming pathogens other than viruses and include dose level of 2.5-10 kGy
- ★ Raduriztion to reduce the number of viable number of viable specific spoilage microbes by radiation dose of 0.75-2.5 kGy.

OR

- ★ Low Dose Application (up to 1kGy)
- ★ Medium Dose Application (1kGy-10kGy)
- ★ High Dose Application (above 10 kGy)

Low Dose Applications (up to 1 kGy)	***Medium Dose Applications (1 kGy to 10 kGy)***	***High Dose Applications (Above 10 kGy)***
Sprout inhibition in bulbs and tubers (0.03-0.15 kGy)	Reduction of spoilage microbes in meat, poultry and seafood under refrigeration 1.50-3.00 kGy	Sterilization of packaged meat, poultry and their products which are shelf stable without refrigeration. 25.00-70.00 kGy
Delay in fruit ripening (0.25-0.75 kGy)	Reduction of pathogenic microbes in fresh and frozen meat, poultry and seafood 3.00-7.00 kGy	Sterilization of Hospital diets 25.00-70.00 kGy
Insect disinfestations including quarantine treatment and elimination of food borne parasites (0.07-1.00 kGy)	Reducing number of microorganisms in spices 10.00 kGy	Product improvement as increased juice yield or improved re-hydration

Labeling of Irradiated Foods

The Food and Drug Administration (FDA) regulates all aspects of irradiation.

FDA requires that both the logo and statement appear on packaged foods, bulk containers of unpackaged foods, on placards at the point of purchase (for fresh produce) and on invoices for irradiated ingredients and products sold to food processors.

Fig 12.1: Radura

Current Status

1. Today 37 countries irradiate food in 170 facilities.
2. There are about 200 large scale radiation sterilization plants operating in USA, Europe and Japan. As on today, half and dozen accelerators facilities are operating in world. Some of them are:

- Odessa Port Facility - In Russia For Wheat
- Radiation Processing Plant - At France for Poultry Meat
- Kasai Electron Beam Irradiation Centre- At Japan for Food and Medical Products
- Hitesys Co. Pilot Irradiation Plant- At Italy For Food Stuffs Polymers,
- In India, Government of India in 1998 has approved irradiation of 14 foods. Two pilot irradiation plants in India (public sector) as per approval of European Commission decision dated 22 March, 2010 are at:

1. Lasalgaon, Nasik for onions.
2. Vashi Navi, Mumbai for spices and herbs.

Private sector radiation plants are (till 2011):

1. Agrosurg Irradiators Pvt Ltd., Mumbai (Maharastra)

2. Hindustan Agro Co operative Ltd., Rahuri Ahmednagar (2011)
3. Agrosurg Irradiators Pvt. Ltd., Mumbai (2009)
4. Innova Agri Biopark Ltd., Bengaluru (2011)
5. Microtrol Sterilization Services Pvt Ltd., Bengaluru (2009)
6. Jhunsons Chemicals Pvt Ltd.Bhiwadi, Rajasthan (2010)

Cost of Irradiated Foods

Irradiation cost may range from Rs 0.25 to 0.50/kg for low dose application such as sprout inhibition and Rs 1-3/kg for high dose application such as treatment of spices for microbial decontamination.

Future Prospects

Radiation processing of food and agricultural commodities can be undertaken both for export and domestic markets. In exports of fruits and vegetables and cut flowers, quarantine barriers can be overcome by using this technology.

Export of meat and meat products can be enhanced since Indian meat has been accepted by many countries. For domestic consumption, radiation processing can be used to facilitate storage, movement and distribution of agricultural commodities from production center to consumption center by preventing post harvest losses during these operations.

There is much scope of irradiated foods for astronauts, cosmonauts and for Naval Forces because of their better hygiene and microbiological safety. The combination of modified atmosphere packaging and irradiation or combination of antimicrobial coatings and irradiation also provide avenues that can potentially extend the usage of irradiation.

15

Related Miscellaneous Facts

Here are some other facts related to basics of science, microbiology and food industry. Knowledge of these facts may help to strengthen the skills of students of Nutrition and Dietetics.

- Joule is the SI unit of heat.
- The study of fungi is known as mycology while the study of algae is known as phycology.
- Father of microbiology is Leeuwenhoek.
- The virus that causes bird flu is H_1N_1.
- In year 1928, Fleming discovered Penicillin, the first antibiotic. While Streptomycin was discovered by Waksman.
- ppm is parts per million means mg/100g or mg/kg.
- Watson and Crick introduced double helical structure of DNA.
- DNA is dexoxyribonucleic acid and has four nucleic acids: adenine, guanine, cytosine and thymine.
- Adenine and guanine are pyrimidine bases while cytosine and thymine are purine.
- RNA is ribonucleic acid and contain uracil instead of thymine as in DNA.
- At -40° reading, Celsius and Fahrenheit is same.

- ★ Father of microbiology is Louis Pasteur.
- ★ Yeast is a kind of unicellular fungus that lack typical mycelia.
- ★ ATA is Alimentary Toxic Aleukia.
- ★ COSTED is Committee on Science and Technology in Developing Countries.
- ★ In BT cotton, BT stands for *Bacillus threogenesis*.
- ★ The virus of bird flu is H_5N_1 which stands for Haematoglutinin 5 neuroamidase 1. Also known as influenza virus.
- ★ β- lactamase, an enzyme is an inhibitor of penicillin.
- ★ Ampicillin is the synthetic identical to pencillin.
- ★ Gramicidin is one of the first antibiotic to be produced commercially.
- ★ Tetracyclin is produced by *Streptomyces griseus*, cephalosporins from *Cephalosporium* and vancomycin from S. *orientalis*.
- ★ Genobiotics are genetically engineered antibiotics.
- ★ Mevinolin is an inhibitor of HMG CoA reductase which is responsible for synthesis of cholesterol.
- ★ Maximum water activity is 1 and relative humidity is 100 percent.
- ★ At lowest water activity (0.60) in honey osmophillic organisms grow.
- ★ Streptokinase is an enzyme which can dissolves clot during heart attack and is obtained from Streptoccocus.
- ★ More acetaldehyde is produced from alcohol, more is hangover.
- ★ ISI stands for Indian Standard Institution.
- ★ International nutrition agencies are WHO, FAO and UNICEF while national nutrition agencies are: ICAR (Indian Council of Agricultural Research), ICMR (Indian Council of Medical Research), NIN (National Institute of Nutrition), FNB (Food and Nutrition Board), CFTRI (Central Food Technological Research Institute), NFI (Nutrition Foundation of India), NNMB (National Nutrition Monitoring Bureau), NSI (Nutrition Society of India), CSWB (Central Social Welfare Board).
- ★ IMTech is Institute of Microbial Technology situated in Chandigarh.

- ★ IICP institute of Crop processing Technology formerly as PPRC is Paddy processing Research Centre.
- ★ NIFTEM is national Institute of Food Technology, Entrepreneurship and Management.
- ★ SNA State Nodal Agency.
- ★ NMPPB is National Meat and Poultry Processing Board.
- ★ APEDA is Agricultural and Processed Food products Export Development Authority.
- ★ MFP is Mega Food Parks.
- ★ IGPB is Indian Grape processing Board.
- ★ Storage structures can be classified as:

Traditional	Modern or improved
Morai	Sacks
Bukhari	Meta or plasic drums
Kothar	Pusa bin
Mud Kothi	Cylindrical grain bin
Muda	Bunker storage
Kanaj	Cement silos
Kuthla	CAP storage
	Sheds

- ★ IMF is Intermediate Moisture Foods. They have 20-50 per cent moisture and 0.60-0.85 water activity. *e.g.* jam, jellies,, honey.
- ★ Biopreservation is a process of preserving food by use of micro-organism and antibiotics.
- ★ Hurdle technology involves using substance or preservation factor to enhance the life of food. It may be physical (high temperature, UV rays, and high pressure), physic chemical (water activity, redox potential, salt, pH) and microbial (antibiotic, bacteriocins).
- ★ Acceptable daily intake (ADI) is the estimate of amount of food additive expressed on body weight basis that can be ingested on daily bases without any appreciable risk to health. It is expressed as mg/kg body weight/day.

Glossary

Acute renal failure: Characterized by rapid onset of renal dysfunction, chiefly oliguria or anuria and sudden increase in the metabolic waste products in blood with consequent development of uraemia.

Acrodermatitis Enteropathica : It is a rare inherited form of zinc deficiency, characterized by periorificial and acral dermatitis, alopecia, and diarrhoea.

Adequate intake (AI): A nutrient recommendation based on observed or experimentally determined approximation of nutrient intake by a group of healthy people when sufficient scientific evidence is not available to calculate an RDA.

Angina pectoris: Commonly known as angina, is chest pain due to lack of blood, oxygen supply and waste removal) generally due to obstruction or spasm of the coronary artries (the heart's blood vessels).

Anions: They carry negative charge e.g. Cl^-, HCO_3^-.

Angular stomatitis : It is inflammation of the mucous membrane of the mouth.

Anorexia: Lack of hunger.

Anorexia nervosa: An eating disorder involving intense and excessive fears of gaining weight coupled with refusal of food to maintain normal body weight

Antabiosis: The type of association in which one organism inhibit other organism.

Anthropometry : The study and technique of taking body measurements.

Antibody: A protein substance produced as a result of antigenic stimulation.

Antibiotic: Chemotherapeutic agent that inhibits the growth of micro organisms.

Antigen: Molecules from a pathogen or foreign organism that provoke a specific immune response.

Apnoea: Cessation of airflow during sleep, preventing air from entering the lungs, caused by an obstruction.

Arithritis: An autoimmune disease in which inflammation of joint occur causing severe to moderate pain.

Ataxia: Loss of control of body movements.

Atherosclerosis: A specific form of arteriosclerosis (thickening & hardening of arterial walls) affecting primarily the intima of large and medium-sized muscular arteries and is characterized by the presence of fibrofatty plaques or atheromas.

Bacteriocins: Bactericidal substances produced by bacteria (protein in nature same as of antibiotics) which act on strains of same or closely related species and are narrow spectrum in comparison to antibiotics.

Balanced diet: Contains different types of foods in such quantities and proportions so that the need for calories, proteins, minerals, vitamins and other nutrients is adequately met and a small provision is made for extra nutrients to withstand short duration of fasting.

Basal metabolic rate: It is the minimum amount of energy that a body requires when lying in physiological and mental rest. If one of these conditions is not met (e.g. shorter time interval for fasting) the measurement is usually termed resting metabolic rate (RMR).

Bioavailability: Percent of the dietary nutrient absorbed and utilized for a specific function by the body.

Biodeterioration: Undesirable changes in properties, chemical composition or structure of material. It may be due to pest, microbes, insects and rodents.

Bradycardia: Low heart rate fewer than sixty beats per minute.

Brix (°B): It is the percentage of sucrose measured by Brix hydrometer.

Candy: A mature fruit/vegetable or its piece impregnated with heavy sugar syrup, till it become transparent and tender.

Cathepsins: These are protein hydrolyzing enzymes that increase tenderness of meat during ageing.

Cation: They carry positive charge *e.g.* Na^+, K^+, Ca^{2+}.

Cerebral Edema: An excess accumulation of water in the brain which can be due to any of a number of causes including excessive water consumption, SIADH, trauma, high altitude exposure, stroke, hypoxia, cardiac arrest, allergic reactions, certain drugs and several other causes.

Chaperons: Heat or cold shock proteins.

Cheilosis : A disorder of the lips marked by scaling and fissures at the corners of the mouth; caused by riboflavin deficiency.

Chylomicrons: Lipoprotein particles that consist of triglycerides (85-92%), phospholipids (6-12%), cholesterol (1-3%) and proteins (1-2%). They transport dietary lipids from the intestines to other locations in the body. Chylomicrons are one of the five major groups of lipoproteins (chylomicrons, VLDL, IDL, LDL and HDL) that enable fats and cholesterol to move within the water-based solution of the bloodstream.

Climacteric fruits: Fruits those which ripen after during storage period. Eg. Mango, banana.

Codex Alimentarius Commission: The Codex Alimentarius Commission, established by FAO and WHO in 1963 develops harmonized international food standards, guidelines and codes of practice to protect the health of the consumers and ensure fair trade practices in the food trade.

Colour: The appearance property or stimulus that result from detection of light after it has interacted with an object.

Colostrum: It is thick, sticky yellowish fluid secreted from the breast of the mother for the first few days after child birth.

Commenslism: An association in which one organism is benefitted and other is neither harmed or benefitted. *e.g. E. coli* in human stomach.

Complementation: Mixing two plant sources together in such a way so that they have complementary effect on each other. *e.g.,* A combination of cereals and pulses as cereals lack lysine and pulses methionine.

Complementary feeding: Feeding the infant with other foods or liquids in addition to breastmilk.

Condiment: A sauce or seasoning added to food to impart a particular flavor or, in some cultures, to complement the dish.

Cordial: A sparkling, clear, sweetend juice from which all pulp and other suspended materials have been removed.

Cori's cycle: Partial oxidation of glucose to lactate in muscle, transporting it to liver for conversion back to glucose and re-supplying to muscles. It has high energy yield than glycolysis.

Crude fat: Fat content of food along with phospholipids, sterols and pigments.

Crude fibre: Proportion of carbohydrates which is in soluble in dilute acid and alkali.

Dehydration: Removal of water by application of artificial heat under controlled conditions of temperature, humidity and airflow.

Delirium : A state in which the thoughts, expressions, and actions are wild, irregular.

Diabetes Inspidus: A condition when urine output & thirst is increased due to insufficient production of antiduieretic hormone (ADH).

Dialysis: A procedure that is a substitute for many of the normal duties of the kidneys. Dialysis can allow individuals to live productive and useful lives, even though their kidneys no longer work adequately.

Diet: Sum of daily/weekly meals considered especially in relation with quality and effect.

Dietary supplement: Also known as food supplement or nutritional supplement, is a preparation intended to supplement the diet and it provides nutrients, such as vitamins, minerals, fiber, fatty acids, or amino acids, that may be missing or may not be consumed in sufficient quantities in a person's diet.

Dietetics: A science that deals with the adequacy of diets during normal life cycle and modifications required during diseased conditions.

Dyslipidemia : Abnormal amounts of lipids and lipoproteins in the blood.

Dyspepsia: Body's inability to digest food.

Dysphagia: Difficulty in swallowing.

Dysnpoea : Difficult respiration.

Eclampsia : A serious complication of pregnancy that includes high blood pressure and excess and rapid weight gain.

Edible film or coating: A thin continuous layer of edible material formed on, placed on/or between the foods or food components. The edible package is an integral part of the food, which can be eaten as a part of the whole food product.

Eicosanoids: Signaling molecules made by oxidation of twenty-carbon essential fatty acids. They exert complex control over many bodily systems, mainly in inflammation or immunity and also act as messengers in the central nervous system. Eicosanoids are derived from either omega-3 or omega-6 EFAs. The amount and balance of these fats in a person's diet will affect the body's eicosanoid-controlled functions, with effects on cardiovascular disease, triglycerides, blood pressure, and arthritis.

Electrolyte: A substance which dissociate into ions when dissolved and can conduct an electric current.

Endospore: A thick walled spore formed around bacterial cell which is resistant to physical and chemical agents due to presence of calcium dipicolinic acid (DPA).

Enzymatic browning: It is a chemical process, involving polyphenol oxidase, catechol oxidase and other enzymes that create melanins and benzoquinone, resulting in a brown color.

Epathy: Lack of interest due to nutrient deficiency.

Ergogenic aids: Substances, devices or practices that enhance an individual's energy use, production or recovery.

Exospores: A heat and desiccation resistant spore formed external to vegetative cell by budding.

Extrusion cooking: A process of high temperature- high pressure short time cooking under controlled conditions of moisture.

Ferritin: A ubiquitous intracellular protein that stores iron and releases it in a controlled fashion.

Food: Anything eaten or drunk that provides nutrition to body.

Food additive: A substance or mixture of substances other than basic food stuff which is present in food as a result of any aspect of production, processing, storage or packaging. It can also be defined as non nutritive substance intentionally added to food to improve flavor, texture and appearance or storage property.

Food adulteration: Addition of non-permitted foreign matter to get more profit by increasing weight or volume of food.

Food exchange lists: Groups of measured foods of the same calorific value, protein, fat and carbohydrate.

Food guide pyramid: Educational tool that shows the dietary guidelines in the easily understood graphic format. It was originally prepared by the Human Nutrition Information Service and published in 1992 by the U.S. Department of Agriculture.

Food infection: In this, organism enters into the tissues of host , it grows and divides there and produces particular symptoms. Symptoms are produced at a slower rate.

Food poisoning: In this, organism grows and divides inside food and produces toxins and if that food is taken immediatel, it causes food poisoning. Symptoms are produced at a faster rate.

Food Science: A study of technical aspects of food. It deals with the study of chemical constituents present in food. OR study of different chemicals present in food, their chemical and physical properties

Food stuff: Any raw substance to be treated as food.

Food technology: Deals with engineering and other scientific and technical issues involved in transforming edible raw material and other ingredients into safe, nutritious and appetizing food product.

Fortification: Addition of one or more than one nutrients in food which may or may not be present in original food. e.g. Vitamin D in milk, Vitamin A and D in fat, Iodine in salt.

Functional foods: Any food that has a positive effect on a person's health, physical performance or state of mind. Eg. Fibrous foods, probiotics, soyabean.

Gelatinization: Starch granule when heated in water, swells and burst.

Genetically modified foods: Produced from genetic modified organisms. Eg yellow rice, BT brinjal.

Harrison Sulcus: Softening of ribs, chest sinks at the line of ribs and cartilages, resulting in broad grooves.

Gluconeogenesis: Process by which glucose is made primarily in the liver by non-carbohydrate carbon substrates such as lactate, glycerol and glucogenic amino acids.

Health: Physical, mental and emotional well being of an individual.

Hematemesis: Presence of blood in vomiting.

Hemodialysis: A procedure in which a machine filters harmful waste and excess salt and fluid from blood. A needle is inserted into arm through a special access point. Blood is then directed through the needle to a machine called a dialyzer, which filters blood a few ounces at a time. The filtered blood returns to body through another needle.

Hemosiderin or haemosiderin: An iron-storage complex. It is always found within cells (as opposed to circulating in blood) and appears to be a complex of ferritin, denatured ferritin and other material.

Hunger: An erg to eat something while, appetite is a desire to eat palatable food.

Hydronephrosis: Distention and dilatation of renal pelvis and calyces leading to atrophy of kidney, caused by obstruction of kidney.

Hydrogenation: A process in which hydrogen react with double bond of unsaturated oil in the presence of nickel as catalyst at high temperature.

Hyperlipidemia: The condition of abnormally elevated levels of any or all lipids and/or lipoproteins in the blood. It is the most common form of dyslipidemia (which also includes any decreased lipid levels).

Hyperplasia: Increase in number of cells.

Hypertrophy: Increase in size of cells, resulting in enlargement of organ.

Hypoxia: Lack of oxygen in body.

Galactosemia: An inborn error of carbohydrate metabolism due to deficiency of enzyme glactose 1 phosphate uridyl transferase because of which galactose is not absorbed.

Gelatinization: The irreversible loss of the crystalline regions in starch granules that occur upon heating in the presence of water.

Gestational diabetes: Condition when pregnant women, who have never had diabetes before, have a high blood glucose level during pregnancy. It may precede development of type 2 diabetes mellitus.

Glomerulonephritis: A term used for diseases that primarily involve the renal glomeruli. It has two types: primary glomerulonephritis and Secondary glomerular diseases.

Gluconeogenesis: Conversion of lactate to glucose or synthesis of glucose from non carbohydrate compounds.

Glycemic index: A ranking of foods based on postprandial blood glucose response compared with reference food.

Goitrogens: Naturally-occurring substances that can interfere with function of the thyroid glandeg. Soyabean-related foods and cruciferous vegetables.

Imbalance: A diseased state that results from disproportion in the nutrient composition of staple diet. *e.g.* Obesity.

Immune Response: Third line of defense. Involves production of antibodies and generation of specialized lymphocytes against specific antigens.

Immunity: Ability of an organism to recognize and defend itself against specific pathogens or antigens.

Infarction: Tissue death caused by a local lack of oxygen due to obstruction of the tissue's blood supply. The resulting lesion is referred to as an infarct.

Irradiation: Controlled application of energy of short wavelength radiation of electromagnetic spectrum (radio waves, micro waves, infra red, visible and UV rays).

Intrinsic factor (IF): Also known as gastric intrinsic factor (GIF). It is a glycoprotein produced by the parietal cells of the stomach. It is necessary for the absorption of vitamin B_{12} in the small intestine.

Ion: Atom or group of atom that carry electric charge.

Iodine number: Number of grams of iodine absorbed by 100 grams of fat. More unsaturation more iodine number.

Jelly: A semi solid product prepared by boiling a clear solution of pectin containg fruit extract.

Kyphosis: Abnormal curvature of spine leading to hump.

Lipotropic factors: Substances that have the ability to remove and prevent fatty deposits in the body. These nutrients essentially perform the task of breaking down and transporting fat from the liver. Lipotropic factors are important because they can help the liver to function better as well as get rid of toxins. They are also needed in order to provide additional energy by burning the transported fat. Eg. carnitine, choline and inositol.

Lathyrism: A paralysing disease of humans (Neurolathyrism) because it affects the nervous system. It is a crippling disease characterised by gradually developing stiff paralysis of the lower limbs occurring mostly in adults who consume the pulse called *Lathyrus sativus* in large quantities.

Maillard reaction: When carbobnyl group of sugars radially combines with the basic amino groups of proteins, peptides and amino acids, it results in sugar amine that cuases browning. *e.g.* Bread crust, roasted meats and roasted beans.

Malnutrition: A diseased condition resulting from relative or absolute deficiency or excess of one or more nutrients.

Marbling: Small streaks of fat that are found within the muscle and can be seen in the meat cut.

Malting: A controlled germination process which activates the enzymes in conversion of cereal starch to fermentable sugars.

Meal: Sum of one or more than one food eaten at one time. *e.g.* Break fast, lunch etc.

Melena: Passage of shiny sticky pitch black stools.

Menu planning: The process of planning and scheduling intake of meals for a general or specific individual requirements.

Mesophilles: Micorganisms that grow best at 25-40°C.

Metabiosis: A type of association in which one organism makes conditions favorable to other organism.

Metabolism: Physical and chemical processes that occurs inside of body that maintain life.

Mutualism: An association in which two organisms are benefitted reciprocally. Eg. Termite and flageller protozoa. Protozoa feed on cellulose taken by termites and termite get their food digested easily.

Myocardial infarction: Death of heart muscle tissues.

Necrosis: Death of cells due to loss of blood supply.

Neutraceutical: A product produced from foods but sold in form of pills powders and other medicinal forms, not generally associated with food and demonstrated to have a physiological benefit or provide protection against chronic diseases.

Nutrients: Chemical constituents present in food.

Nutrient density : Amount of nutrients for a given volume of food.

Neutrogenomics: A study of effect of foods and food constituents on gene expression.

Nutrition: A branch of science that deals with ingestion, digestion, transportation and absorption of nutrients and finally excretion of waste in form of feces and urine.

Nutrient requirement: The minimum amount of the absorbed nutrient that is necessary for maintaining the normal physiological functions of the body.

Nutritional status: Condition of health as effected by food eaten by body.

Obligatory nitrogen losses: As protein intake declines, the efficiency with which amino acids are reutilized increases, but small amounts of amino acids are continuously degraded and their nitrogen is lost from the body, even when no protein is being consumed.

Osteoblasts: Mononucleate cells that are responsible for bone formation,

Osteoclast: A type of bone cell that removes bone tissue by removing its mineralized matrix and breaking up the organic bone. This process is known as bone resorption.

Osteoid: Unmineralized organic portion of the bone matrix that forms prior to the maturation of bone tissue.

Organic foods: Foods that are produced without using conventional pesticides, fertilizers and ionizing radiations.

Over nutrition: A disease state that result from consumption of excessive amount of food over an extended period of time.

Oxytocin : A posterior pituitary gland hormone that stimulates the uterus to contract during child birth and the breasts to release milk.

Peptide Bond : Linkage between carboxyl group and amino group

Peritoneal dialysis: Blood is filtered inside the body after the abdomen is filled with a special cleaning solution.

Phagocytes: White blood cells that protect the body by ingesting (phagocytosing) harmful foreign particles, bacteria and dead or dying cells.

Phagocytosis: A mechanism by which single cells of the animal kingdom such as smaller protozoa, engulf and carry particles into the cytoplasm.

Phenol: Aromatic compounds with hydroxyl groups offers resistance to diseases and pests in case of plants.

Phytochemicals: They are certain organic components of plants that are beneficial to human health in a different way from traditional antioxidants. They are sometimes referred to as phytonutrients, but unlike the traditional nutrients (protein, fat, vitamins, minerals), they are not "essential" for life so the term phytochemical is more accurate.

Pickle: Preservation of food in common salt or in vinegar.

Pocker's back: Calcification of ligaments of spine.

Polycythemia: Increased number of RBCs due to excess of cobalt.

Polydypsia: Increased in thirst.

Polyphasia: It is excessive hunger.

Polyurea : It is excessive urination..

Prolactin : A pituitary hormone that stimulates and maintains the secretion of milk.

Prostaglandins: They are like hormones as chemical messengers. They do not move to other sites, but work right within the cells where they are synthesized. They are biochemically synthesized from the fatty acids.

Portal Hypertension: An increase in the blood pressure within a system of veins called the portal venous system. Normally, veins coming from the stomach, intestine spleen and pancreas merge into the portal vein, which then branches into smaller vessels and travels through the liver. If the vessels in the liver are blocked, it is hard for the blood to flow, causing high pressure in the portal system.

Prebiotic: A non- digestible food ingredient that stimulates growth or activity of beneficial bacteria in the digestive system. *e.g.* dietary fibre, curd and milk.

Preservation: Process of prevention of decay of foods, thus allowing it to be stored in a fit condition for future use.

Probiotics: Micro-organisms that impose beneficial effect on human health during their presence in human gut. *e.g. Lactobacillus acidophilus* and *Lactobacillus brevis*.

PDCAAS: Protein Digestibility Corrected Amino Acid Score.

Psychrophiles: Organisms responsible for spoilage at low temperature (0°C or low). They cannot grow at >25°C, optimum is 15°C.

Psychrotrophs: Micro organisms that grow at 0°C. Also known as facultative psychrophiles. *e.g. Clostridium botulinm* and *S. aureus*.

Pulmonary Circulation: The circulation that pumps deoxygenated blood from the right side of the heart via the right ventricle into the lungs to obtain oxygen.

Pulmonary Hypertension: It It is far less a common type of high blood pressure that affects specifically the arteries in the lungs. Pressure in the lung arteries is normally significantly lower than the pressures in the systemic circulation. Pulmonary hypertension occurs when the pressure in the pulmonary circulation becomes abnormally elevated. It is a serious condition that becomes progressively worse and eventually proves fatal.

Putrefaction: Anaerobic decomposition of proteins that causes foul odour in food due to hydrogen sulhide, methyl and ethyl sulphites and amines.

Radiation: Emission and propagation of energy through space or medium.

Radappertization: A method of radiation sterilization using radiation dose of 30-40KGy.

Radicidation: A method of radiation sterilization using radiation dose of 2.5-10KGy. Used for milk.

Radio isotopes: Unstable isotopes of elements. They are the atoms of same element with different mass number but same atom number.

Radurization: A method of radiation sterilization using radiation dose of 0.75-2.5KGy. Used for meat and fruits.

Recommended Dietary Allowance (RDA): Levels of intake of essential nutrients considered to be adequate to meet the known nutritional needs of all healthy persons. It is also used as a guide to determine the nutritional adequacy of individual diets. RDA= Requirement of nutrient+ margin of safety .

Recombinant DNA technology: The process of formation of new genotype by resortment of genes by exchange of genetic material. It requires restriction endonuclease to cut double stranded DNA and ligases to join parts of two different DNA.

Reference man: A man between 20-39 years with weight 60kgs who works for 8 hours, spends 8 hours in bed, 4-6 hours in sitting and 2 hours in walking.

Reference woman: A woman between 20-39 years with weight 50kgs who works for 8 hours, spends 8 hours in bed, 4-6 hours in sitting and 2 hours in walking.

Restoration: Restoring nutrients removed during processing of food. e.g. Adding thiamine and iron to flour which has lost during milling.

Sauerkraut: A clean, wholesome product with characteristic flavor, obtained by shredded cabbage fermentation in 2-3% salt and requires no starter culture.

Sclerosis: Increased density and hyper calcification of bones of spine, pelvis and limbs.

Scoliosis: A person with scoliosis has a sideways curve to their spine. The curve is often S-shaped or C-shaped.

SGOT: Serum Glutamic Oxaloacetic Transminase.

SGPT: Serum Glutamic Pyruvic Transaminase.

Sherbet or syrup: A clear sugar syrup which is artificially flavoured.

Sideroblastic anaemia: A disease in which the bone marrow produces ringed sideroblasts rather than healthy red blood cells.

Smoking point: When fat is heated above boiling point of water, it begin to smoke, flash and burn.

Specific deficiency: A pathological condition resulting from relative or complete absence of an individual nutrient of the staple diet. Eg. Goitre.

Spice: A dried seed, fruit, root, bark, or vegetative substance used in nutritionally insignificant quantities as a food additive for flavor, color or as a preservative that kills harmful bacteria or prevents their growth.

Spoilage: A change undesirable in nature that makes food unfit for consumption Or decay or decomposition undesirable in nature.

Sponification number: Number of milligrams of KOH reqired to sponify 1 g of fat or oil.

Squash: Consists of strained juice containing moderate quantities of fruit pulp to which sugar is added for sweetening.

Supplementation: Providing additional nutrients to diet through addition of other food. Eg soya flour in wheat flour.

Starvation: A condition with zero intake of food.

Symbiosis: A type of relationship in which two organisms grow with each other and both are benefitted. *e.g.* Ruminants and bacteria.

Synbiotics: Mixture of prebiotics that beneficially affect the host by improving the survival and implantation of live microbial dietary supplement in gastrointestinal tract.

Systemic Circulation: In the human body, the circulation that pumps oxygenated blood from the left side of the heart via the left ventricle to all parts of the body.

Syntrophism: An association in which two organisms depend on each other for nutrients and growth factors.

Tannins: Polyphenolic compounds. They can be hydrolysable (made of polyhydric alcohol) and condensed (flavanols).

Teratogenic : Substances such as chemicals of radiations that cause abnormal development of an embryo.

Thermophiles: Micorganisms that grow best at 45°C. Optimum is 55°C. *e.g. Bacillus, Clostridium.*

Thermoduric micorganisms: Mesophiles who can survive pasteurization temperature. *e.g. Lactococcus* and *Lactobacillus.*

Toxemia : An abnormal condition of pregnancy characterized by hypertension, edema and proteinuria.

Transamination: A process in which non essential amino acids can be formed by pyridoxal phosphate and alpha ketoglutaric acids.

Transferrins: Iron-binding blood plasma glycoproteins that control the level of free iron in biological fluids

Transthyretin: A serum and cerebrospinal fluid carrier of the thyroid hormone, thyroxine (T4) and retinol binding protein bound to retinol.

Total energy expenditure (TEE): Energy spent on average in a 24-hour period by an individual or a group of individuals.

Total Soluble Solids (TSS): It is amount of sugar and soluble minerals present in fruits and vegetables.

Type 1 diabetes: Results from the body's failure to produce insulin and requires the person to inject insulin. (Also referred to as insulin-dependent diabetes mellitus, IDDM for short and juvenile diabetes.)

Type 2 diabetes: Results from insulin resistance, a condition in which cells fail to use insulin properly, sometimes combined with an absolute insulin deficiency. (Formerly referred to as non-insulin-dependent diabetes mellitus, NIDDM for short and adult-onset diabetes.)

Ulcerative colitis: Inflammation and ulceration of colon, resulting in frequent passage of stools with blood and mucus. Nitric oxide can be a cause of it.

Under nutrition: A pathological condition resulting from consumption of inadequate amount of food for over an extended period of time. *e.g.* Marasmus.

Upper intake level : The highest continuing level of daily nutrient intake that is likely to pose no risk of adverse health effects in almost all individuals, in the specified life stage group.

Vitamins: Organic compounds required in the diet in small amounts to perform specific biological functions for normal maintenance of optimum growth and health of the organism. Our body cannot not make them on its own. Thus, we must obtain them from the foods we eat or via vitamin supplements

Viral hepatitis: Liver inflammation due to a viral infection. It may present in acute or chronic forms.

VO_2 max: A measure of a muscle's maximal capacity to use oxygen in microlitres of oxygen consumed per gram of muscle per hour.

Xenobiotic: A chemical found in an organism, not produced or expected to be present in it. e.g. Antibiotic in humans.

24 hour recall : A retrospective dietary assessment method that analyzes each food and drink consumed over the previous 24 hours.

Appendix

Some Important Formulas

1. Resting Energy Expenditure

Males- 10 x wt in kgs+ 6.25xht in cms-5 x age +5

Females- 10 x wt in kgs+ 6.25xht in cms-5 x age -161

Then multiply it by activity factor

1.5- women and 1.5 for men

2. Mastuda'a Formula

0-10 per cent TSB= HBEE x1.5

10-30 per cent TSB= HBEE x1.5

>30 = HBEE x1.7

Where TSB is Total Surface of Body and HBEE is Harris Benedict Energy Expenditure

3. Harris Benedict Equation

It is a method used to estimate an individual's basal metabolic rate (BMR) and daily calorie requirements.

BMR- In men: 66.5+ (13.8 x W) + (5xH) – (6.8 x A)

BMR- in women: 655.1+ (9.6 x W) + (1.8 x H) – (4.7 x A)

Then BMR is multiplied by activity factor

where A is for Age, W for Weight and H for Height.

4. Curreri Formula

Adults- 25 kcal x pre burn wt + (40x per cent TSB)

Children 30-100 kcal x pre burn wt + (40x per cent TSB).

5. Body Metabolic Rate (BMR) can be calculated by

$$\frac{\text{Activity}}{24 \times 60 \times \text{body weight}}$$

6. Respiratory Quotient

$$\frac{CO_2 \text{ exhaled}}{O_2 \text{ consumed}}$$

7. Protein Efficiency Ratio (PER)

$$\frac{\text{Gain in weight (g)}}{\text{Protein intake (g)}}$$

8. Digestibility Coefficient (DC)

$$\frac{100 \times \text{N intake- endogenous fecal N}}{\text{N intake}}$$

Or

$$\frac{100 \text{ X In-(Fn-Fe)}}{\text{In}}$$

where Fn is fecal nitrogen on protein diet and Fe is fecal nitrogen from protein free diet

9. Biological Value (BV)

$$\frac{\text{N digested- N lost in metabolism X 100}}{\text{N digested}}$$

$$= \frac{\text{In-(Fe-Fn)-(Un-Ue) X 100}}{\text{In-(Fe-Fn)}}$$

where Fn is fecal nitrogen on protein diet, Fe is fecal nitrogen from protein free diet, Un is urine nitrogen on protein diet, Ue is urine nitrogen on protein free diet and In is total nitrogen intake

10. Net Protein Utilization (NPU)

$$\frac{\text{DC X BV}}{100}$$

$$\frac{\text{Body Nitrogen of test group} - \text{Body Nitrogen of non-protein group} + \text{Nitrogen consumed by non-protein group}}{\text{N consumed by test group}} \times 100$$

11. Net Protein Ratio

$$\frac{\text{Gain in weight of test group - loss in weight of non protein group}}{\text{Protein intake}}$$

12. Chemical Score

$$\frac{\text{Content of most limited amino acid in test protein}}{\text{Content of same amino acid in egg in egg protein}}$$

13. Protein Digestibility Corrected Amino Acid Score (PDCAAS) in percentage

$$\frac{\text{mg of limiting amino acid in 1g of test protein} \times \text{fecal true digestibility (per cent) X 100}}{\text{mg of same amino acid in 1g of reference protein}}$$

14. True Digestibility

$$\frac{\text{Absorbed nitrogen}}{\text{Nitrogen intake}} \text{X 100}$$

15. Na Conversion Formula

mg of Na x 2.5= Mg of salt

mg of salt x 0.4=Mg of Na

16. Broka's Index

Height in cms-100= ideal body weight.

17. Water Activity

$$\frac{P}{P_o}$$

i.e. ratio of vapour pressure of solution to vapour pressure of solvent.

18. Relative Humidity

100 x water activity (aw)

19. The Rf Vale in Chromatography

$$\frac{\text{Distance of solute from origin}}{\text{Distance of solvent from origin}}$$

20. Normality (Acid)

$$\frac{\text{Desired normality x Desired Volume x Equivalent weight x 100}}{\text{Percent purity x Specific gravity x 1000}}$$

21. Normality (Base)

$$\frac{\text{Desired normality x Desired Volume x Equivalent weight}}{1000}$$

22. Molarity

$$\frac{\text{Desired molarity x Desired volume x molecular weight}}{1000}$$

23. mM

$$\frac{\text{Desired normality x Desired volume x molecular weight}}{1000 \times 1000}$$

24. ppm

$$\frac{\text{Mass of component x } 10^6}{\text{Total mass of solution}}$$

25. Water Activity (a_w)

$$\frac{\text{Vapour pressure of water in food}}{\text{Vapour pressure of food at same temperature}}$$

26. The Formula of Calculating Energy by Alcohol (Proof/2 = per cent alcohol)

0.8 X proof X ml = Kcal.

27. Iodine Value

$$\frac{\text{grams of iodine absorbed}}{\text{100g of fats}}$$

28. Grain strength is used to express the percent of acetic acid in vinegar. One grain strength = 0.1 per cent acetic acid.

References

Antia FP and Abraham P. 2002. Clinical Nutrition and Dietetics. Oxford University Press, India. Inc. p 390.

Bamji Mehtab S et al. (ed). 1998. Textbook of Human Nutrition. Oxford and IBH Publishing Co. Pvt Ltd, New Delhi.

Frazier WC and West Hoff DC. 1986. Food Microbiology. Tata McGraw Hill Publishing Co. Ltd, New Delhi.

Garrow JS and James WPT. 1993. Human Nutrition and Dietetics. Churchill Living Stone.

Gopalan C, Rama Sastri BV and Balasubramaniam SC. 1996. Reprinted. Nutritive Value of Indian Foods. NIN, Hyderabad.

Manay S and Shadaksharaswamy M.1987. Foods-Facts and Principles. New Age International (P) Publishers Ltd, Chennai.

Shubhangani A Joshi. 2002. Nutrition and Dietetics. Tata Mc.Graw Hill Publishing Co. Ltd, New Delhi.

Srilakshmi B. 2002. Nutrition Science. New Age International (P) Publishers Ltd, Chennai.

Srilakshmi B. 2003. Food Science. New Age International (P) Publishers Ltd, Chennai.

Mudambi SR and Rao SM. 1989. Food Science. New Age International (P) Publishers Ltd, Chennai.

Mudambi SR and Rajagopal MV. 2012. Fundamentals of Food, Nutrition and Diet Therapy.6th Edition. New Age International Ltd. Publisher, New Delhi

Swaminathan M. 1988. Essentials of Food and Nutrition, Volume I and II. The Bangalore Printing and

Publishing Co. Ltd, Bangalore.

Recommended Dietary Allowances for Vitamins, NIN, ICMR, Hyderabad, India (2010) and Vitamin and Mineral Requirements in Human Nutrition, FAO/ WHO (2004) (http://www.who.int/vmnis/indicators/haemoglobin).

www.nutrition.org

www.micronutrient.org

http://www.public.navy.mil/bupers-npc/support/physical/Documents/Guide per cent 204-%20Body%20Composition%20Assessment%20(BCA).pdf

www.ncbi.nlm.nih.gov/pubmed/20176521

ww.elmhurst.edu/~chm/vchembook/555prostagland.html

http://www.nhlbi.nih.gov/health/health-topics/topics/atherosclerosis/

http://www.traceelements.com/Docs/Nutrient%20Interrelationships-Minerals-Vitamins-Endocrines.pdf

www.ingramcontent.com/pod-product-compliance
Ingram Content Group UK Ltd.
Pitfield, Milton Keynes, MK11 3LW, UK
UKHW021954270726
14060UKWH00002B/507

9 789388 173063